The Old Moon and the New

The Old Moon
& The New

V A Firsoff

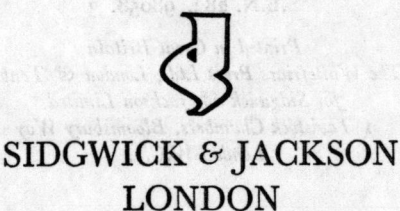

SIDGWICK & JACKSON
LONDON

First Published 1969
Copyright © 1969 V. A. Firsoff
Copyright © in the postscript 1969 Patrick Moore

S.B.N. 283. 98058. 3

Printed in Great Britain
by The Whitefriars Press Ltd., London & Tonbridge
for Sidgwick & Jackson Limited
1 Tavistock Chambers, Bloomsbury Way
London W.C.1.

Contents

List of Plates

List of Text Figures

Introduction: between the past and the future

The dreaming of dreams is a pleasant pastime, but few things are more disturbing than dreams beginning to come true. This is about to happen to man's ancient dream of setting foot on the Moon. What will it reveal? The unexpected or the commonplace?

In any event we are faced with a great historical moment, the importance of which cannot be overestimated, however much the keen edge of expectation may have been honed to bluntness by an excess of publicity. It is a time to take stock of our knowledge, the more so as a few hours' sojourn on the Moon's surface, however fruitful, cannot solve all the problems and nullify all the previous results. This is why I have entitled the present book *The Old Moon and the New*. But there is more to it than that; it also forms a kind of sequel to my earlier work, *Strange World of the Moon*, subtitled *An Enquiry into Lunar Physics*, published in Britain in 1959[7] and set in type before the first *Lunik* hit the Moon and the headlines as the harbinger of the new era.

Thus, by chance rather than foresight, that other book, too, came to mark an important scientific divide. Its position there, however, was not wholly a matter of timing, but, as the later developments have shown, also of treatment and outlook. Indeed, the view of the Moon it presented was in many respects closer to 1969 than to 1959.

History is not erased, it is only overlaid by later happenings, and it may be instructive at this juncture to retrace our steps to that earlier date and earlier book.

It had been preceded by a prolonged period of doldrums in lunar studies, when the Moon became the favourite preserve of the amateur astronomer and was almost wholly abandoned by the profession, except for a few specialized aspects of traditional fundamental astronomy, such as lunar motions or physical librations, which required both meticulous patience and advanced mathematical treatment, and theories propounded by people who had little personal experience of the Moon at the eyepiece. There was a measure of conflict between such theories and the observational data amassed mainly by amateur observers, who were often highly skilled and painstaking, but lacked the prestige of professional standing.[12]

Such considerations have little to do with science, whose essence lies in the impartial pursuit of truth, but they have a good deal to do with the way the human world ticks. Having had professional training, but not being professionally employed, I was uncommitted and unclassifiable in this context, and this is a coign of vantage for breaking loose from the mental ruts.

Be this as it may, it became clear to me when the war ended that some spectacular developments in space flight could not be long in coming (though the reality has far exceeded my modest expectations) and would betoken a fundamental revolution in the methods and content of astronomy and in our ideas in general, a point that is still lost on many with their eyes too close to the ground.

Since the Moon would inevitably be the first target of space exploration, I decided to devote most of my attention to it, both by direct telescopic study and by extending my knowledge of the theoretical and practical work done by others.

To begin with, all I had at my disposal were various items of war surplus equipment, of high optical quality, but not designed for astronomical purposes. From such elements I contrived a $2\frac{1}{2}$-inch refractor with a viewing attachment of microscopic design, whose relatively weak lenses could be moved to and fro, to give graded magnification. The only meaningful observations such an instrument allowed were colour estimates extended to the whole or large portions of the lunar disk. For this purpose I used a Dufay tricolour separation set, whose filters isolated from the total light approximately the three basic responses of the colour-sensing cones in the retina. These filters were mounted in a rotary carrier, which I patented and an independent version of which has lately become known as the *Moon-Blink* (P. K. Sartory). The carrier was placed between the objective and the ocular of the microscopic attachment, and both were subsequently used to good effect with a $6\frac{1}{2}$-inch and 12-inch Newtonian reflectors. Indeed, I found this eyepiece superior to most conventional types.

The colour effects observed by me were first reported in the May 1958 issue of *Sky and Telescope*,[6] and most of them are amply confirmed by the recent infrared-ultra-violet comparison photography carried out by E. A. Whitaker at the Lunar and Planetary Laboratory, Tucson, Arizona.[17] One notable exception is my blue-violet observations of occultations, where the star appeared to fade gradually at the dark limb of a young crescent Moon,[3] which nobody has troubled to check up.

Though not altogether irrelevant, this is a side issue. My main task lay in the sifting out of the available evidence, in order to arrive at a coherent picture of the physical condition of the lunar surface, which inevitably involved the formative selenological processes upon and beneath it.

The inherited view of the situation existed in two main variants, which may be described as 'official' or perhaps 'doctrinal', and 'observational' or perhaps 'amateur', respectively. In the first the Moon was an inert, substantially unchanging body, buffeted by interplanetary matter, large and small, and obligingly preserving its primordial condition for the edification of human posterity. Its surface was a meteoritic impact counter, and all (well, nearly all) of its features were due to this cause and very ancient.

Opposed to this were the ideas of many systematic observers, among whom the Harvard professor W. H. Pickering occupied a somewhat extreme position.[11, 12] They saw evidence of change, variable colourings, glows and obscurations, some kind of action that was not all due to external causes. Some lunar formations had a worn-down ancient appearance, while others looked clear-cut and fresh, bearing witness to erosion and the continued operation of internal, *enselenic,* forces, broadly similar to the igneous activity of our geology.

Both these approaches may be said to have been fathered by Robert Hooke's *Micrographia,* published in 1667. Hooke observed craters formed by bullets dropped into pipeclay slurry, as well as those arising from the bursting bubbles in a boiling watery mash of powdered alabaster. Thence he inferred that lunar craters could have been formed either by impact or by internal heat, and favoured the latter explanation, largely because the existence of meteorites was not recognized in his lifetime. Indeed, his idea of a subsiding bubble comes very close to the modern conception of the collapse caldera, where the dome, uplifted in the overlying strata by an intrusion of liquid magma, caves in when its support is withdrawn through draining away, shrinkage due to the escape of volatiles, or to eruptive evisceration.

The new information brought by the orbital and ground probes that are whipped off to the Moon every few months, if not weeks, has modified the outlook in many respects. Thus, for instance, the deep dust hypothesis is no longer tenable. Many old observations out of joint with the official doctrine, which were once dismissed as illusions of incompetent amateurs, have now been given professional accolade. But the age-long controversy over the origin of lunar ring mountains and all that it implies remains unresolved and unabated.

To return, however, to my private story, it was, of course, perfectly clear that some meteorites must have hit the Moon as they had the Earth, leaving marks of impact upon their surfaces. Yet on the Earth these marks were few, far between and relatively insignificant, the more so as many alleged *astroblemes* were shown by competent geologists to be *maars* (rimless volcanic craters) or cryptovolcanic depressions. Thus recognizing the reality of meteoritic bombardment and the better preservation of its effects on the surface of the Moon was not the same as jumping to the conclusion that this was the main, let alone the only, process that had shaped its features. I will admit that the idea was repugnant to me philosophically *a priori* as another example of geocentrism, desperately striving to preserve a unique position for man's home planet. Nevertheless, such promptings apart (and they cannot be dismissed either), the detailed study of lunar formations disclosed to my eye the work of familiar geological forces, only modified by the consequences of reduced planetary mass.

I was forcibly struck, as were some other selenographers and selenologists, and the American geologist J. E. Spurr[14] in particular, by the existence on the Moon of a global pattern of tectonic lineaments, involving the most varied types of surface feature, with hardly an exception. This stamped them as functional links in a single chain of cause and effect, arising from internal forces of the lunar globe and the tidal action of the Earth, rather than from any purely haphazard agency, such as a barrage of cosmic artillery. I, therefore, proceeded to study and map these lineaments under varying illuminations, this work being eventually compounded in a large three-colour map appended to my *Moon Atlas* in 1961.[9]

My views on this subject, however, found their first published expression in a short paper 'On the structure and origin of lunar surface features', which appeared in the *Journal of the British Astronomical Association, 66* (8), in 1956,[4] as part of a debate with Gilbert Fielder, then still under the influence of the prevalent meteoritic idea, though our views came to converge more and more in the course of time. My first provisional map of the lunar fracture grids was included in another paper, coincident with his on the same subject, almost exactly a year later in the same *Journal, 67* (8/2).[1, 5]

Ancillary to my investigation of the lunar grids was spherical projection photography, or rectified photography as it has become known since. The original idea was due to F. E. Wright and dated back to 1925 when a 'Committee on the Study of Physical Features of the Surface of the Moon' was set up under him in Washington.

He projected negatives of the Moon on to 12-inch chemical flasks coated with photographic emulsion over a distance of 135 feet in an underground tunnel of the Mount Wilson Observatory. This made it possible to obtain an image of the Moon where most effects of foreshortening were eliminated and the features near the limb were restored to their correct relative positions on the lunar globe, which facilitated their study, interpretation and mapping. The method, however, proved to be cumbersome and difficult, and only a few of these 'Wright's spheres' had been made before the Committee was wound up in 1940.[13]

Working on a total budget of about £10, I followed a different procedure of projecting positive lunar slides on to a whitewashed geographical globe and photographing the latter at suitable angles. This, of course, did not result in a permanent spherical record, but the photographs thus obtained were highly instructive and, it seems, superior to the original Wright's spheres in limb detail. A selection of them was included in *Moon Atlas,* published by

Hutchinson of London in 1961 and in the following year by the Viking Press of New York.[9] I started my work early in 1958, with the knowledge of Ewen A. Whitaker, then at Greenwich, and it seems that similar experiments were embarked upon at Yerkes at about the same time and were eventually to lead to the publication in 1964 of the *Rectified Atlas of the Moon* by Whitaker, Kuiper *et al.*, prepared at the Lunar and Planetary Laboratory of the University of Arizona.[16]

In the main, however, my thoughts on lunar grids had been inspired by Spurr, although he had mentioned specifically only 'polar grids'. Many of the lines mapped by me have been rediscovered later on by various investigators, with what seems to be an unnecessary duplication of effort, but some of the master fractures, such as the great off-latitudinal crack traversing Mare Nubium and Oceanus Procellarum, still await recognition.

I can claim no originality in holding that the main mechanism of formation of the lunar ring mountains was intrusion by magma or similar incompetent rock, such as ice, causing a domical uplift, followed by the withdrawal of support and collapse, producing a ringed depression, exemplified by terrestrial volcanic calderas. This interpretation, at least in its igneous version, had been proposed long before I sat down to write *Strange World of the Moon*. What, however, had greatly impressed me in this connection was the insufficient attention given to the consequences of the low lunar gravity, coupled with the virtual absence of subaerial denudation, if not of erosion.

There was nothing new in the idea that the Moon may be coated with pumice, but the full extent of the vesiculation and swelling of any lava poured out under vacuum conditions, as revealed, for instance, in E. W. Washburn's experiments with molten glass[17] (a similar siliceous material), and the general ineffectiveness of compacting forces on the lunar surface remained unrecognized. Only recently has it become usual to think of lunar surface rocks as being lighter than water and suitable experiments have been carried out at the National Physical Laboratory at the instance of the University of London Observatory and Fielder in particular. Yet the thought that lunar mountains may be, to quote from *Strange World of the Moon,* 'as insubstantial as meringues' has not percolated yet to the official levels. The layer of rock froth continues to be generally regarded as at most some metres thick, with a familiar terrestrial type of rock, such as granite – or is it dunite? – underneath.

This, however, seems to me a basic misunderstanding of the past history of the Moon, stemming from the tendency to shy away from continuous processes in favour of some kind of 'big bang' that can be conveniently dated. Such an approach is mathematically handy : the process can be numerically estimated, put into equations, computerized and a neat solution found. The only small snag about such solutions is that they are untrue. So, too, in this case. Flows of lava or fluidized ash, perhaps of mud and water, did not occur just once, but over and over again through the long succession of 'geological' ages; and since there were no effective agencies to remove the resulting deposits, the entire surface of the Moon will consist of such lightly compacted rocks to a depth of kilometres (it is unfashionable to speak of miles these days!). Eventually gravitational compression, aided by the action of heat and water, will prevail, and they will assume a denser texture, but with a gravity of $1/6$ g acting on materials whose bulk density is only a fraction that of water this will not happen a few metres below the surface, nor will it result in anything like our granite.

These ideas, first proposed in *Strange World of the Moon,* will be developed more fully in later chapters, where they properly belong.

That earlier book also dealt at some length with the effect of the permafrost seal in lunar subsoil, in which it anticipates Professor Thomas Gold by at least a year, the probable existence of subsurface ice and underground hydrosphere, due, on the one hand, to the low density of the surface formations making water a relatively heavy liquid with a tendency to sink to lower levels, and, on the other, to its being kept away from the surface by permafrost. The idea that lunar vulcanism may be comparable to our geysers, fumaroles or mud volcanoes is traceable to the writings of Pickering[11] and Spurr,[14] although the latter's approach to selenology is distinctly in the 'hell-fire' tradition, but I believe I am right in saying that its first tentative application to the genesis of the maria is to be found in Chapter 7, entitled 'The Sunless Sea', of *Strange World of the Moon.*

It is not often that my thoughts concur with those of Professor

H. C. Urey, who is one of the extreme advocates of the meteoritic hypothesis, seeking in the Moon clues to the past of the Solar System. But, writing in 1966,[15] he says:

... I have interpreted the lava flows as coming from Mare Tranquillitatis when the mare was produced by a great collision during the early history of the Moon. In recent years because of several lines of evidence that have led Professor Gold particularly to suggest water as an important feature of the Moon and which have led me for quite different reasons to make similar suggestions, I have thought that perhaps it was a water flow of some kind instead of a lava flow ...

'The great collision' and reference to Professor Gold apart, this sounds very much like an echo of 'The Sunless Sea', which goes to show that its ideas have not lost their vitality even in an impact context.

Another line of thought that is still lying fallow and was first broached in *Strange World of the Moon* is the importance of gas sorption to the surface regime of planets of small mass. Indeed, the only instance I know of where this matter has been taken up is Victor Kachur's (Westinghouse Astronuclear Laboratory, Pittsburg, Pa.) paper, *Evidence of CO_2 Sorption Effects on the Surface of Mars* (1967) (a preview copy). Unlike Urey, Kachur expressly refers to my work (for, regrettably, there exists nothing else to refer to in this connection). So far as the Moon is concerned, however, the field is still virgin.

The main lines of selenological thought initiated in *Strange World of the Moon* were developed and set forth in succinct form in a further book, entitled *Surface of the Moon* and published in 1961 (Hutchinson),[8] which is still in print and being referred to.[2, 10] Nevertheless, the intervening seven years have produced more in terms of lunar exploration than the preceding centuries since Galileo first directed his 'optick tube' at the queen of our nights and saw the pockmarks on her face. Indeed, the mass of new material is bewildering in its richness and decrees a searching re-examination of many favoured viewpoints. Yet mammoth

Plate 1. The astronomer's old traditional Moon (south at the top). A near-full Moon (13.8 days old) photographed by Chappell with the 36-inch refractor of the Lick Observatory. *Lick Observatory.*

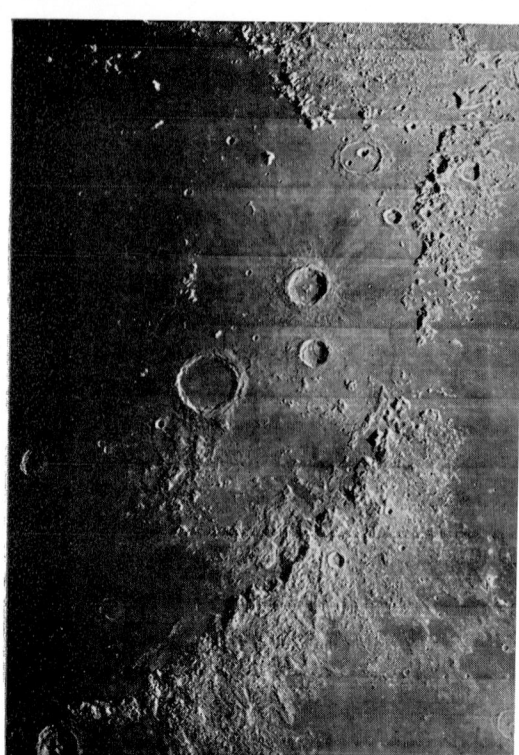

Plate 2. Mare Imbrium with its mountain girdle (north at the top). Moving counter-clockwise from bottom left corner, the mountains are: Apennines (note the Conon and Hadley Rilles), Caucasus and Alps, showing movement of crustal blocks (p. 94). A portion of the *Orbiter* photograph IV–120M₁.

N.A.S.A.

Plate 3a. An oblique view across the crater Copernicus northwards to the Carpathians, taken by *Orbiter II* from 45.5 km above the surface.

Plate 3b. The same part of the northern walls of Copernicus seen vertically from above. Note annular step faults, lava flows, structure of the rocks, roughness of the floor, and stream beds. V–156M. (*Orbiter* frames.)

N.A.S.A.

Plate 4a. A sharp-ridged elementary ring structure, Gambart (I–134M).

Plate 4b. A well-developed young crater (Phillips), showing subsidence along multiple ring fractures and numerous cumulo-domes on the floor, which is substantially devoid of craterlets (V–37M).

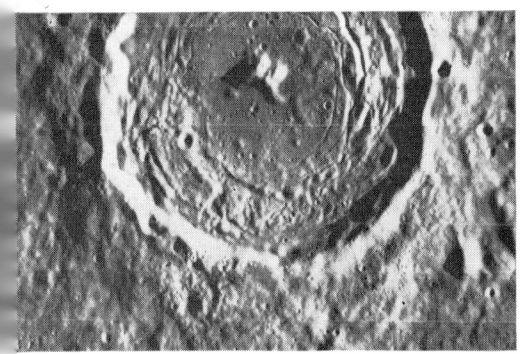

Plate 4c. An older crater structure, Taruntius with subdued walls, open ring fractures in the interior, fewer domes and numerous craterlets (I–31M).

N.A.S.A.

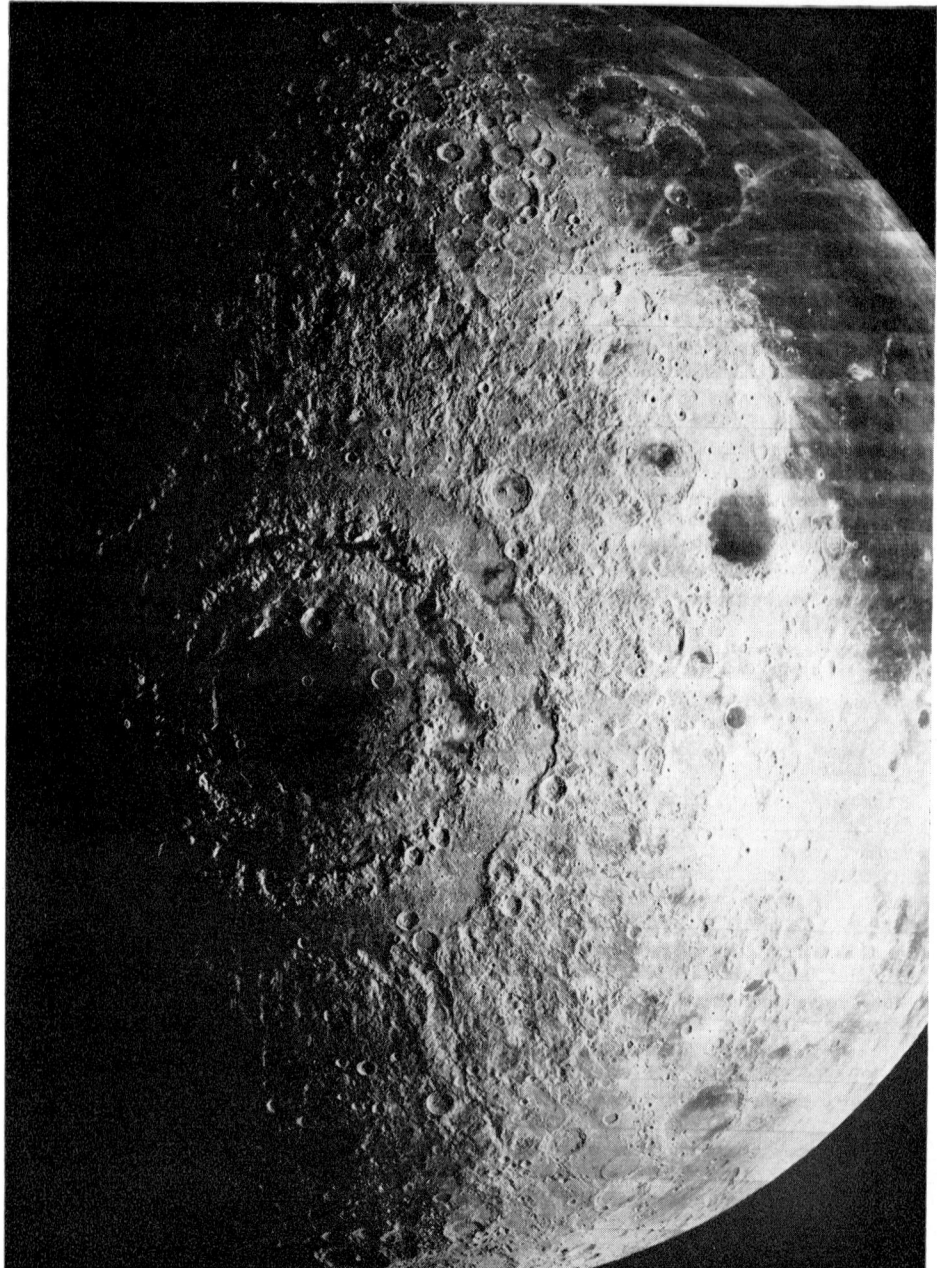

Plate 5. The western hemisphere of the Moon with Mare Orientale and Oceanus Procellarum on the right above. Note the mountain rings and the numerous lava flows around them. Inghirami and Wargentin (Plate 13a) are at the bottom on the right, Riccioli (Plate 18), is the large ring to the north-west of the dark 'eye' of Grimaldi (IV–187M). *N.A.S.A.*

organizations, such as the N.A.S.A., are notoriously slow movers, and, seeing that so much of what I have thought and said before 1959 has been borne out by the new age of lunar research, it would seem that there is still room for a 'lean dog, a keen dog, a bad dog, a mad dog that bays the moon, to keep fat souls from sleep'.

Familiarity may not breed contempt except where deserved, but continuous familiarity certainly blunts the keen edge of perception. This was why, after my earlier concentrated work on the Moon, I had decided to give it comparative rest and moved on to other subjects. I was also biding my time, to see what lunar probes would reveal, until the moment when further data would bring quantitative rather than qualitative extensions of the information already to hand. This moment appears to have arrived.

It may be questioned – and I have been inclined to do so myself – if such an *apologia pro vita sua* is really called for to preface a scientific work. On the other hand, the personal scientific history of a book is not irrelevant to its contents, and such expressions as 'it is thought', 'it has been shown', to say nothing of 'the present writer', are only awkward circumlocutions for the first person singular that serve further to aggravate the tortured English in which scientific works are so often composed.

A wholly impersonal and even dull recital of facts may be justifiable in a technical paper, addressed to other workers in the same field who are keenly interested in the subject. But is this treatment right for a book? What audience is it to be addressed to, a university lecture theatre or a wider assembly? I suspect that even experts are less willing to swallow page upon page of dry-as-dust expertise than they would usually admit. Sheets of mathematical reasoning are followed step by step only by the specialist concerned with the problem in question; other readers tend to skip them and, taking the result for granted, say that 'the author *has shown* this or that' (which is often enough far from the case).

I have, therefore, followed my usual plan of seeking a wider public by writing as readably and accessibly as possible without becoming trivial or too elementary, and relegating the more specialized stuff into the appendix.

With this, for what it may be worth, *accipe posteritas*!

REFERENCES

1 FIELDER, GILBERT (1957). 'The lunar grid system', *J.B.A.A.*, **67** (8/2).

2 FIELDER, GILBERT (1965). *Lunar Geology*. Lutterworth Press, London.

3 FIRSOFF, V. A. (1956a). 'Lunar occultations observed in blue light', *J.B.A.A.*, **66** (7).

4 FIRSOFF, V. A. (1956b). 'On the structure and origin of lunar surface features', *J.B.A.A.*, **66** (8).

5 FIRSOFF, V. A. (1957). 'On the tectonic grids of the Moon', *J.B.A.A.*, **67** (8/2).

6 FIRSOFF, V. A. (1958). 'Color on the Moon', *Sky and Telescope, 17* (7).

7 FIRSOFF, V. A. (1959). *Strange World of the Moon*. Hutchinson, London, and 1960, Basic Books, New York.

8 FIRSOFF, V. A. (1961a). *Surface of the Moon*. Hutchinson, London.

9 FIRSOFF, V. A. (1961b). *Moon Atlas*. Hutchinson, London, and 1962, Viking Press, New York.

10 MOORE, PATRICK and CATTERMOLE, PETER (1967). *The Craters of the Moon*. Lutterworth Press, London.

11 PICKERING, W. H. (1903). *The Moon*. Doubleday, Page & Co., New York.

12 PICKERING, W. H. (1916). 'Lunar Changes', *The 7th Report of the Section for the Observation of the Moon of the B.A.A.* London.

13 SPRADLEY, L. H. (1962). 'Lunar Globe Photography', *Communications of the Lunar and Planetary Laboratory*, No. 6. University of Arizona.

14 SPURR, J. E. (1944–1949). *Geology Applied to Selenology*, Vols. I–IV. Science Press, Rumford Press and Business Press.

15 UREY, H. C. (1967). 'Study of the Ranger pictures of the Moon' in 'A Discussion on the physics of the Moon and its environment', *Proc. R.S., Series A*, **296** (1446). London.

16 WHITAKER, E. A. *et al.* (1963). *Rectified Lunar Atlas*.

17 WHITAKER, E. A. (1966). 'The Surface of the Moon' in *The Nature of the Lunar Surface, Proceedings of the 1965 IAU-NASA Symposium*. The Johns Hopkins Press, Baltimore.

Abbreviations used

J.B.A.A. = Journal of the British Astronomical Association.
B.A.A. = British Astronomical Association.
Proc. R.S. = Proceedings of the Royal Society.

2

The double planet

Reality is a continuum, a system of inter-relationships where nothing stands alone. It is literally true that one cannot move a finger without stirring the stars in their courses, but *fortunately* – and I say so advisedly – the effect on the stars is unobservable by any conceivable technique. For any science, even speech itself, would become impossible if we could not meaningfully isolate some aspects of this continuum and consider them as if they were unrelated to the rest. But it is just as well to realize that this process of abstraction involves an inevitable distortion of the truth. If the departure from the truth is no greater than in neglecting the gravitational response of stellar masses to the movement of a finger – well and good. Very often, however, this distortion is quite considerable and may reduce a complicated mathematical reasoning to a meaningless piece of academic showmanship.

To come down to Earth and Moon, the latter is not just a decoration on the tapestry of heaven. The relationship between our home planet and its companion world is manifold and close.

Silently, invisibly the Moon drives the Severn Bore, which is, of course, but one example of the ubiquitous tides.

Broadly speaking, tides are due to the drop in the gravitational attraction of the Moon over the diameter of the Earth. This attraction is inversely proportional to the squared distance of the attracting mass, which for large distances and in the first approximation may be taken to be concentrated at a point – the centre of gravity or barycentre. At first the decrease in gravity, which is the acceleration due to the attractive force and so provides a measure of it, is very rapid – the gravitational field is steep. If, starting at 2,000 miles from the barycentre, we move away to 4,000 miles from it, the distance is doubled and the gravity reduced by four. But if we similarly add 2,000 miles to a distance of 10,000 miles, the proportional increase in distance is only by a factor of $1\cdot2$ and entails a drop in gravity by a factor of $(1\cdot2)^2 = 1\cdot44$; the field has become much flatter.

The Sun is roughly 93 million miles and the Moon 240,000 miles away. This is why, although the Sun amounts to 27,069,000 Moons in mass and so in gravitational force, the solar tide is only about a third as strong as the lunar.

Tides cause not only a movement of fluids on or above the ground, but also a distortion of the solid body of the Earth, which tends to bulge out on the opposite sides of the Earth-Moon line. This would be exactly true if the Moon were stationary in our skies, like a synchronous communication satellite. In actual fact, however, the Moon moves only through some 2° of the circle while the Earth turns once round. The Earth is not a perfectly elastic body, so that the tidal bulge raised by the Moon takes some time to subside as well as to arise, owing to which it is carried ahead of the Earth-Moon line by the Earth's axial spin and the Moon tries to pull it back into alignment. This generates a decelerating torque, braking the rotation of the Earth and lengthening the day. But the attraction between the bulge and the Moon is mutual and equal both ways, and the bulge tends to pull the Moon forward in its orbit, as a result of which the Moon is speeded up and recedes from the Earth.

This traffic in angular momentum, which measures the 'quantity of rotation' and is expressed by the product of momentum (mass × velocity) into the radius vector (distance from the centre of revolution), between the Earth and the Moon will continue until the tidal bulge raised in the Earth by the Moon is immobilized at the sublunar point, as has already happened to the Earth-induced bulge in the Moon, which has come permanently to face the Earth. Apart from minor wobbles, called librations, of which more later, the Earth hangs motionless in the lunar sky, and the Moon would be similarly brought to apparent standstill in our sky with the Earth-bulge immobilized at the sublunar point. This means that the terrestrial day and the lunar orbital period have become equal.

Sir Harold Jeffreys has calculated that this will require about 50 æons, or 50 thousand million years, by when the mean distance of the Moon from the Earth will have increased from 240,000 to 340,000 miles and the Moon's orbital period from 27 to 35 days, the latter being also the length of the terrestrial day.[1]

Even this, though, is not the end of the story, as the solar tidal torque will continue to eat up the rotation of the Earth, acting counter to the gravitational attraction of the Moon, which will now be losing its orbital momentum, to keep the bulge at the sub-lunar point against the Sun, and be drawn closer to the Earth once more.

I will not pursue this cosmic *Ragnarok* to its catastrophic conclusion, as the whole situation is a paper abstraction. For one thing, according to R. A. Lyttleton, the Moon may yet escape from the Earth.[9] For another there are further forces at work. Thus, Eric Holmberg contends that the diurnal atmospheric tide, due not to solar gravitation, but to the heat of sunrays, supplies energy to the Earth's spin. Expanding by day and contracting by night, the atmosphere acts as a 'heat engine'. Since, further, the period of resonance, in which the atmosphere would oscillate if excited into vibration, is approximately 24 hours, the 'heat engine' will tend to stabilize the day at this length.[13]

H. C. Urey[14] pursued a different line of thought.

Rotation is a physical quantity comparable to heat, which is molecular motion; both obey the law of conservation of momentum, except that rotation can be readily turned into heat through friction, but heat cannot be turned into rotation with equal ease (it is turned into rotation in a steam engine). If a hot body is compressed it grows hotter, because the same amount of heat is concentrated in a smaller volume. Similarly, if a rotating body is compressed it will spin faster. The speed of rotation may thus be likened to temperature and angular momentum to heat. In a closed system the sum of angular momenta of parts is constant, bar friction, and cannot be altered without the intervention of an external force; but through the interaction between parts angular momentum can be passed from one part to another.

At birth, materials of various specific weights may be taken to have been mixed more or less uniformly within the body of the Earth, but in the course of time its interior would become heated by radioactive decay and melting would ensue (see p. 20), the heavier matter sinking towards the centre and the lighter rising towards the surface. The net result of the operation would be an increased concentration of mass towards the centre, paralleled by a concentration of angular momentum at positions close to the axis of revolution, which by the law of conservation must give faster rotation. Urey has calculated that this mechanism could have reduced the day from about 30 hours to its present length.

Any shrinkage or expansion of either the lunar or the terrestrial globe – and both have been mooted and may have occurred – would likewise affect their rates of spin, even though it can be shown that the Moon's axial angular momentum could not contribute much to the evolution of its orbit.[8] Less drastic changes in surface loading, arising from large subsidences, mountain-building processes or the growth of ice sheets, may cause polar wandering and react upon the length of the day.[5, 6] The interaction between the magnetic field of the Earth and the solar wind may also have to be taken into account. In other words, the situation is very complicated and it is difficult to be certain how much is due to what cause.

Nevertheless, Munk and MacDonald have found astronomical evidence for a secular lengthening of our day attributed to lunar tides, by 1·8 seconds per 100,000 years.[7] Less certainly, it has been claimed from the diurnal skeletal ridges in Devonian corals that there were 400 days to the year 400 million years ago. This would correspond to a higher rate of increase in the length of the day, amounting to about 2·4 seconds per 100,000 years if it is assumed that the year has not changed as well. The tidal evolution of the Earth-Moon system, however, involves the recession of the Moon from the Earth and so a weakening of the tides, which would have been stronger in the past.

Since, as we shall see later on, the plane of the Moon's orbit is inclined to that of the terrestrial equator at 23·4°, the rotational displacement of the tidal bulge of the Earth does not follow the Moon's orbital motion and the attractive force is skewed, which reacts on the shape and inclination of the lunar orbit.

According to Kopal[7] the semi-major axis of the lunar orbit would now be increasing by about 3·2 cm per year, and its inclination to the ecliptic declining very slowly by 1·9 seconds of arc per million years, with a simultaneous increase in eccentricity of 0·12 per æon. The accuracy of these figures is suspect, but some such changes are taking place, and one thing is beyond doubt: the Moon was closer to the Earth in geological past and its action correspondingly stronger.

To revert to the present tense, the atmospheric tide due to lunar gravity is very slight, owing to the fact that our air ocean is shore-less. Indeed, observations made at the Royal Greenwich Observatory between 1854 and 1917 have revealed a fluctuation in barometric pressure associated with the lunation (the period within which the Moon repeats its phases) equal to barely 0·01 mm Hg or torr. This is a negligible fraction not only of the total mean sea-level pressure of 760 torr, but of the day-to-day pressure fluctuations involved in weather changes. Yet it is logical to surmise that in regions which, unlike the Thames estuary, are enclosed by high mountains, confining the lowermost and densest portion of the atmosphere, the lunar tide will be much more marked, although no investigation of this kind has been carried out to date so far as I know. On the other hand, statistical correlations between rainfall averages and lunar phase in certain areas, such as the Mediter-ranean basin, have been claimed.

Owing to the Earth's rotation about the barycentre of the Earth-Moon system, in which the Earth counter-dances to the Moon in pantographic reduction by a factor of 0·0123 (ratio of the Moon's to the Earth's mass), the Earth is about 6,000 miles closer to the Sun at full than at new Moon, and this entails a corresponding change in the heat of sunshine. But a more likely cause of the weather-phase correlation may be sought in the action of the Moon on the ionized layers, where lunar tides of up to 40 miles in ampli-tude have been observed. Space probes have registered no mag-netic field in the vicinity of the Moon, but by intercepting the solar wind the Moon will create a magnetic wake, to which the effect is probably due. An electric field may likewise be involved.

In folklore the Moon is invested not only with an influence on the weather, but with a more-or-less magical power over the growth of plants. Experiments made to test this belief have given inconclusive results. Nevertheless, it cannot be denied that both through the agency of the tides and of moonlit nights the Moon has had and is having appreciable biological repercussions. In the subtler field of mental environment the Moon has been prominent in early religious belief, legend and story, as well as in the practical sphere of time reckoning and the development of ideas about heavenly bodies and natural laws, to take as an example Newton's law of universal gravitation, which he tested on the motions of the Moon. In the days to come the Moon has an important part to play as an air-free observational base and a stepping stone to space, with the consequences that will be clear in retrospect, but are not easy to foresee.

Although, to borrow a simile from Walter Shepherd, in point of rigidity a steel ball bearing is a better small-scale model of the Earth than the traditional apple, we have seen that the effect of tidal forces is not limited to the soft parts of the Earth's anatomy. It extends to the crustal and subcrustal layers, and under high pressures solid rocks creep slowly when exposed to sustained stress. Such solid or *rheid* flow is an important geological concept,[5] which is expressed in terms of time as a ratio of viscosity to rigidity. This is a concealed measure of speed, as the time is the interval required for the flow to traverse a given distance, usually one centimetre.

The rate of rheid flow increases with temperature and when melting point is reached it changes to ordinary viscous flow. The crustal formations are very rigid and their rheid flow is some thousand million times slower than that of ice in a glacier (asssuming equal stress); but the subcrustal layers of the mantle, about 20 miles below the surface, are in a semi-molten plastic state and have a rate of flow only about 10,000 times slower than that of glacier ice, which is quite adequate for them to perform the task of a responsive resilient medium on which the crust is buoyed. If the crustal load is increased by sedimentation the medium yields to it and subsides; similarly it redresses itself when the overburden is lighted up by erosion, so that the land surface is upheaved. This behaviour is called *isostasy*.

Now a correlation is suspected between lunar phases, and so tides, and earthquakes, which are usually caused by the slipping of rock strata along crustal fractures or faults. It is indeed logical to expect such sensitive geological lineaments to respond to the stresses due to the tidal fluctuation in lunar attraction and/or the inertial reaction of the rock masses to the Earth's rotation about the common barycentre of the Earth-Moon system (see p. 15), as well as to the variation in surface loading due to the movement of coastal waters. Thus the Moon must inevitably affect terrestrial tectonics, but the rhythmical stresses generated by tidal forces should also provoke rheid movements in the crustal formations and especially below the crust. The continents and the roots of the light sialic rock at the base of the mountain ranges would be bobbing up and down imperceptibly upon the heavier glassy basalt of the mantle as the Moon moves round the sky. However slight, these repetitive displacements would have a cumulative effect and may generate regular convection cells, shifting land and sea in the coarse of geological ages.

Although the existence of subcrustal convection is generally admitted by geologists, they tend to think of the Earth as an isolated body rather than as a member of the Earth-Moon system, and, to say the least, the latter aspect requires investigation. In sum total there can be no doubt that the history of our planet and of our species would have been different in the absence of the Moon.

The relationship is, of course, reciprocal, and, since the Earth is equal to 81·30 Moons in mass, the direct gravitational attraction exerted by the Earth on the Moon is this much stronger. We have seen, however, that tidal forces depend not on gravity alone, but on its drop over the diameter of the body exposed to it.

Now the mean distance between the Earth and the Moon is 384,400 km (238,900 miles), the Earth's equatorial diameter measures 12,756 km = 7,927 miles, and that of the Moon 3,480 km = 2,160 miles. A simple calculation will show that the mean drop in lunar gravity between the sublunar and antilunar point is about 6.5 p.c., while the corresponding decrease in terrestrial gravity over the diameter of the Moon is only 2 p.c. This makes the force of the terrestrial tide on the Moon 25 times stronger than that of the lunar tide agitating our seas. The effect would be terrific if the Moon were in free rotation and had seas comparable to ours.

And this is precisely why the Moon, as well as all the other inner satellites of the Solar System, have tied-up or synchronous rotation in which the axial and the orbital periods coincide, so that, except for orbital eccentricity, tidal stresses are eliminated. The Moon's orbit has an appreciable eccentricity of 0·0549, a point to which we shall return. Meanwhile it is of some interest to consider how effective the tidal action of the Earth would be in reducing the independent spin of the Moon to tied-up condition.

We do not know if the Moon has ever rotated asynchronously. Assuming, however, that it were in independent rotation the tidal interaction at its present level would be eating up this rotation some 10,000 times faster than it is that of the Earth (MacDonald).[10] Indeed, Jeffreys' figure is 17,000.[1] Thus on the present reckoning, if we start off with any arbitrarily short axial period, this would be brought to synchronism within 10-20 million years, a time that is very short by geological standards. Moreover, since a simultaneous recession of the Moon is involved and tidal action is approximately proportional to the 6th power of the orbital semi-major axis, the rate of decrease in the axial spin must have been very much faster in the past, and the whole episode relatively brief.

On the other hand, seeing that the Moon is moving away from the Earth, the lunar month must have been lengthening steadily. The whole process of tidal evolution may be followed backwards, as was first done by G. H. Darwin in 1880, to its logical conclusion when the Earth and the Moon were in contact and revolving about their common centre of gravity in approximately 4 hours.[1, 7]

This gave rise to the idea that the Moon was born out of a molten Earth through the tidal action of the Sun and the Pacific Ocean is the scar left by the departing Moon.

In the classical treatment of the problem, solar tides acting on the rapidly rotating body, close to the point of gravitational instability, are amplified by resonance and throw it into pulsation of increasing amplitude, in which it assumes the shape of a triaxial ellipsoid, rather like an egg. Eventually the pulsating mass breaks up into two unequal parts: the larger Earth and the smaller Moon.[11] The proposition has been investigated by F. Nölke and more recently by R. A. Lyttleton with a negative result.

Writing in 1935, Sir Harold Jeffreys[6] has summed up the situation thus:

> The friction at the boundary of the core would dissipate the energy of the tide so rapidly that it could never reach an amplitude more than about 1/20 of the radius which would be far too small to give instability; and it seems, according to Nölke,

that the velocity of the detached mass would be so small that it would at once fall back into the primary.

There is also the difficulty that a large satellite in hydrostatic equilibrium could not exist within Roche's limit, which falls at 2·89 Earth radii and where the shearing force of the tides would shatter it into a ring, somewhat like those of Saturn.

This point of view is generally accepted today.[9] Yet the trouble with mathematical investigations of this kind, fascinating as they are, is that they deal with artificial problems created at the outset by postulating the initial conditions and forces at work. In these, of course, the numerical data derived from observation are taken into account, but the question arises if the information fed into the formulæ is exhaustive.

In the present case it may be necessary to consider thermal, as well as gravitational instability. An important factor is the age of the Earth. This is currently put at $4\frac{1}{2}$ æons, mainly on the basis of the radioactive dating of the oldest crustal rocks accessible to examination. Whether these rocks are coeval with the Earth, however, is another matter. In the geological cycle, which is in constant operation, igneous rocks – and these are mainly granitic in the crust – decay, to be reformed as sedimentary formations, which are overlaid by later sediments, become metamorphosed and eventually restored to igneous constitution. We are unable to say how many such cycles the Earth's crust may have gone through. On the other hand, the radioactive ages of meteorites, which, too, are members of the Solar System, extend to as much as 10 æons, and an opinion is gaining strength that the Earth may be much older than has been thought.

Radioactive elements, responsible for the heat of planetary interiors (at least in the terrestrial planets), decay with time, and their initial proportions in the nascent Earth must be calculated backwards from the present abundances. Uranium 1 has a half-life of 4·6 æons. Thus on the present reckoning there will have been about twice as much of it when the Earth was born. But if the age of the Earth is doubled the initial amount of uranium 1 is quadrupled, and the effect on the elements with shorter half-lives is much more drastic.

G. P. Kuiper[8] has calculated on the basis of the quoted age of the Earth and of the Solar System that the liberation of energy by radioactivity was originally ten times higher than now, and all bodies in excess of 100 km in diameter must have experienced central melting. This has been, it is true, disputed by Urey,[14, 15] but the latter's assumptions are a little strained, and Kuiper's conclusions are probably accurate enough.

It is thought that the planets arose through accumulation of cold material and only subsequently became heated from inside outwards by gravitational compression and radioactive decay. If their age is substantially above the $4\frac{1}{2}$ æons assumed by Kuiper, however, radioactive heating could have been very rapid, leading to catastrophic expansion or explosion. In a swiftly revolving body this could result in a ring of matter being cast off at the equator, as in Laplace's old genesis of the planetary system. This ring would be magnetically coupled to the main mass, as proposed by H. Alfvén, which would, on the one hand, check its outward movement and, on the other, remove it beyond Roche's limit at the expense of the primary's angular momentum. The rotation of the planet would be slowed down and its stability increased, no further satellites being born.

There are indications that the inner satellites of the Jovian planets have originated in this way.[8] But has the Moon?

For one thing the Jovians exceed the mass of the Earth between 14·52 (Uranus) and 317·89 (Jupiter) times, and even now their internal temperatures may be over $100,000°C$: with the escape of radiation blocked by an opaque envelope the possibility of nuclear fusion cannot be discounted at the centre of Jupiter and Saturn. For another the Moon is very large in relation to the Earth, their diameters comparing as 0·2725 to 1, and masses – we shall recall – as 1 to 81·30; while relatively the largest Jovian satellite Triton, which is anomalous withal, is 750 times less massive than its primary Neptune, their diameter ratio being 1:12. In fact, Triton's diameter is estimated at 3,700 km, which is not much above the Moon's 3,480 km, although Triton's mean density of 5·1 puts it in a class with the Earth (5·52) or Mercury (5·41) rather than the Moon (3·34) and gives it a correspondingly higher mass. The diameter of Triton has been repeatedly revised in recent years and may be revised again, but the point is that if Neptune itself were a companion of Jupiter it would still fail to reproduce the scale of the Earth-Moon system, and the actual largest moon of Jupiter, Ganymede, boasts a diameter of only 5,000 km and a mass 1/12,000 of the primary's.

Thus the Earth-Moon system is a double planet rather than a planet with a normal satellite. Furthermore, either of the suggested modes of origin should have placed the nascent satellite substantially in the plane of the planet's equator. This requirement is met by the inner moons of the Jovians, except Triton; but the orbit of our Moon subtends an angle of 23·4° with the equatorial plane of the Earth. Both this orbit and the axis of the Moon keep constant inclinations to the ecliptic, which defines the plane passing through the centre of the Sun, with no reference to the Earth whatever.

Indeed, solar and terrestrial gravity draw level at a mean distance of 161,800 miles (260,300 km) from the Earth's centre, which the Moon never approaches closer than 221,463 miles (365,500 km), so that the Sun's pull on the Moon is stronger than the Earth's. This is also why the path of the Moon as seen from the Sun is always concave towards it. This path may, therefore, be described as a planetary orbit perturbed by the Earth. Since, however, the Moon falls within the *activity sphere* of the Earth, which has a mean radius of 576,700 miles[2] and within which the gradient of the terrestrial field is steeper than that of the solar, terrestrial perturbations exceed the solar in magnitude. Nevertheless, the Earth's hold on the Moon is none too secure : all elements of the lunar orbit are subject to considerable solar and planetary perturbations, and it has already been mentioned (p. 13) that the Moon may yet escape and become an independent planet.

It is, therefore, reasonable to assume that it originally was independent and was captured by the Earth at a relatively late stage of development. It would thus be, as I have put it in *Strange World of the Moon,* not 'the Earth's fair child, but a foundling'[3] (with apologies to copycats).

Indeed, Darwin's calculations were based on the assumption that the Earth behaved like a globe of viscous liquid, and when the problem was re-examined by H. Gerstenkorn (1955) and G. J. F. MacDonald (1964) for an elastic Earth a very different picture emerged. There are minor differences between them, but in both analyses the eccentricity and the inclination of the lunar orbit increase backwards in time, leading to a parabolic or hyperbolic orbit, possibly retrograde, with an obliquity of over 45° and the point of closest approach at (Gerstenkorn) or within (2·74 Earth-radii, MacDonald) Roche's limit, when the Earth turned once round in 5–6 hours. A parabolic or hyperbolic orbit in relation to the Earth implies capture.[4, 7, 10]

This would have occurred 1·4 æons ago according to Gerstenkorn[4] and 1·75 æons ago according to MacDonald,[10] and so after the geologically established origin of life on the Earth. The event would have been catastrophic for both planets. The lunar tides raised on the Earth would have been some 8,000 times stronger than today, and so (Kopal) attained 'amplitudes comparable with the present mean depth of the oceans'.[7] This is an interesting point, for intense earthquakes and global igneous activity of unimaginable violence appear inevitable in these conditions, and if any higher life had evolved on the Earth during the preceding ages it would certainly have perished, only bacteria and low marine organisms being able to survive. The records of pre-Cambrian rocks are too scanty for either confirming or disproving this conclusion, but may well be worth investigating from this angle.

If the Moon had entered the Roche limit at least its surface would have been shattered, and it has been suggested by Alfvén (1963) that the Moon may subsequently have been bombarded with its own fragments. Large-scale melting seems inevitable. Some of the angular momentum would thus have been dissipated by friction and conversion to heat, though Gerstenkorn does not think this mechanism would have been sufficient to convert a close approach into capture, the conditions for which have been shown by R. A. Lyttleton to be very stringent. Radial tides in the body of the Moon seem, however, to satisfy the requirements by keeping the angular momentum constant with a strong decrease in orbital energy. Gerstenkorn concludes that by their means even an hyperbolic orbit could have been converted into an ellipse of equal parameter in a single encounter.[4]

MacDonald[10] writes (1967): 'If the Moon ever approached within 2·7 present Earth radii, then the surface ellipticity would have been 1 part in 10. Such a large surface flattening would certainly have led to a discernible distortion of the present crater shapes. It thus appears that the Moon was either not within about 5 radii of the Earth, or that the craters date from a time recent compared with the time of close approach.' In actual fact, old craters within and near the tropical region of the Moon show signs of east-west crustal shortening, attributable to the drawing-in of an equatorial bulge. This effect, which J. E. Spurr[12] has called *appresion*, will be considered in a further chapter, and it does seem to amount to 1 part in 10 or so. Appresion is absent in post-Ptolemaic craters.

To sum up, while much remains obscure, the Moon most probably was an independent planet until $1\frac{1}{2}$–2 æons ago.

REFERENCES

1 BALDWIN, R. B. (1949). *The Face of the Moon*. University of Chicago Press.

2 EHRICKE, K. A. (1960). *Space Flight, Vol. I. Environment and Celestial Mechanics*. Van Nostrand, New York.

3 FIRSOFF, V. A. (1959). *Strange World of the Moon*. Hutchinson, London.

4 GERSTENKORN, HORST (1967). 'The importance of tidal friction for the early history of the Moon'. *Proc. R.S., Series A.* **296** (1446).

5 HOLMES, ARTHUR (1965). *Principles of Physical Geology*. Nelson, Edinburgh and London.

6 JEFFREYS, HAROLD (1935). *Earthquakes and Mountains*. Methuen, London.

7 KOPAL, ZDENEK (1966). *An Introduction to the Study of the Moon*. Reidel, Dordrecht.

8 KUIPER, G. P. (1954). 'On the origin of the lunar surface features', *P.N.A.S.*, **40** (12).

9 LYTTLETON, R. A. (1967). 'Dynamical capture of the Moon by the Earth', *Proc. R.S., Series A.* **296** (1446).

10 MacDONALD, G. J. F. (1967). 'Evidence from the surface configuration of the Moon on its dynamical evolution', *Proc. R.S., Series A.* **296** (1446).

11 RUSSELL, N. H. (1935). *The Solar System and Its Origin*. New York.

12 SPURR, J. E. (1949). *The Shrunken Moon*. Business Press.

13 *The Observatory* (1956). 'Geophysical Discussion'. **76** (892).

14 UREY, H. C. (1952). *The Planets*. Oxford University Press.

15 UREY, H. C. (1955). 'Some criticisms of *On the Origin of the Lunar Surface Features* by G. P. Kuiper', *P.N.A.S.*, 41.

3

Motions of the Moon

Change is exciting, stability – reassuring, and astronomy caters for both, for while the events and changes it portrays are of awesome grandeur, the chances of, say, being hit by a stray meteorite or of witnessing a nearby nova explosion are vanishingly small, and the usual pace of change is so slow by the standards of human history, let alone individual life, as to create an impression of order, permanence and peace. The constellations, the planets, the Moon were there centuries before we were born and will be there long long after we are gone, very much as they are now.

In the foregoing chapter we have reviewed the Earth-Moon system as a double planet in its evolutionary aspect, noted the insecurity of the Earth's grip on its companion, and caught some glimpses of possible catastrophic developments in the past and future. This discussion necessitated reference to the orbits, motions and figures of the two planets as the functional variables of their relationship, but was unsuited to give a clear picture of the here-and-now, which it is our present object to tidy up and work out in more detail.

As seen from the Sun, the Moon's path is always concave towards it, but with reference to the Earth the Moon describes an ellipse that does not greatly depart from a circle and has a variable eccentricity, averaging 0·0549. As Kepler's laws demand, the centre of revolution is at one of the foci of this ellipse, which by Newton's law of universal gravitation is the common barycentre of the system, and we will recall that the Earth, too, revolves about it. Its path is an exact replica of the lunar orbit reduced by a factor of 0·0123, in inverse ratio to the interacting masses. This puts the focus at 2,940 miles = 4,721 km from the Earth's centre, and so well within the 'terrestrial ball'.

The resulting motion of the Earth causes the nearest celestial bodies, such as Venus, Mars and some eccentric asteroids, of which Eros is the most important, measurably to shift position among the stars. From such displacements the mass of the Moon can be obtained in terms of the Earth's mass if the distance of the observed body is known. An intra-Martian asteroid is particularly handy for this purpose, for not only may it approach the Earth closer than the neighbouring planets, but its telescopic image is substantially a stellar point of light, so that the parallactic shift is both larger and can be measured with greater accuracy. Nowadays a radar echo offers a quick and easy way of determining the distance, which used to be a source of trouble and uncertainty, while artificial lunar satellites afford a direct means of studying the gravitational field of the Moon with a degree of accuracy beyond the dreams of classical astronomy.

Since the Earth is also rotating about its polar axis, whose inclination to the normal to the line joining the centres of the two bodies varies from 0 to 28° 36′ 44″, the lunar motion of the Earth generates a variable centrifugal force, superimposed on the radial tides due to the drop in lunar gravity over its diameter (p. 12).

The distance of the Moon from the centre of the Earth averages 384,400 km = 238,885 miles, but at the closest point of the orbit, called *perigee*, it drops to 364,400 km = 221,463 miles, and rises at the farthest point, or *apogee,* to 406,730 km = 252,710 miles. This entails a corresponding variation in the reciprocal tidal action, and, more directly, a proportional one in the apparent diameter of the Moon. Observationally the latter is of no great consequence, except in one respect. The mean apparent diameter of the Moon in seconds of arc is 1865·16 and that of the Sun 1919·26. The two are very nearly equal, but on the average the Moon is just a little too small fully to eclipse the Sun. The result is an annular eclipse, where the dark disk of the Moon is ringed with solar brilliance. But both diameters vary sufficiently, owing to the orbital eccentricities of the Earth and the Moon, to make total eclipses possible.

In fact, the apparent diameter of the Moon is 204.8″ larger at perigee than at apogee.

As referred to the stars, the Moon completes one circuit of its orbit on the average in 27 days 7 hours 43 minutes and 11.5 seconds, or if you are a stickler for astronomical precision, 27.321661 mean solar days. This is its *mean sidereal period*. Since, however, the Moon and the Earth move jointly round the Sun, the latter, as seen from the Moon, falls behind the fixed stars by about 1/13 of the full circle at every revolution, the illumination of the lunar globe follows suit and the phases recur after 29 days 12 hours 44 minutes and 2.9 seconds. This is the *synodic period* of the Moon, coincident with the *lunation*, the first measuring the return of a celestial body to the same position relatively to the Sun, and the second the time between two successive new Moons.

The sidereal period varies by as much as 7 hours owing to the solar and planetary perturbations, which impart the same variation to the synodic period, additionally affected by the ellipticities of the Earth's and the Moon's orbits, the first altering the apparent motion of the Sun, both owing to its changing distance and the fluctuation in the orbital velocity of the Earth (this is inversely proportional to the square root of the distance from the Sun, or properly the radius vector), and the second by reason of the uneven progress of the Moon about the Earth. All this adds meaning to the qualification *mean*.

We will recall that the Moon rotates synchronously, so that its mean orbital period is also its true *sidereal axial period*, which, though, is constant to within one part in 30 million (T. Banachiewicz, 1950). By the same token the lunation closely corresponds to the solar day of the Moon, reckoned from sunrise to sunrise, or sunset to sunset; but the two are again non-coincidently affected by perturbations, librations, arising from the uniformity of axial rotation and the variability of orbital velocity (to be considered later on), and the inclination of the polar axis of the Moon.

Certain regularities, however, are discernible among these bewildering complications, and these were first formulated in three empirical laws by J. D. Cassini in 1693 :

1. The Moon rotates uniformly about an axis fixed within the lunar globe and its period of rotation is equal to the sidereal period of the Moon's orbital revolution.

2. The rotational axis of the Moon makes a constant angle with the normal to the plane of the ecliptic passing through the centre of the Moon.

3. The normal to the plane of the ecliptic and the normal to the plane of the lunar orbit drawn through the centre of the Earth define a plane containing the polar axis of the Moon.

These laws are only approximate and remained unexplained until 1764, when Lagrange won the award offered by the Paris Academy of Sciences for solving the dynamical problem involved in them. They can be formulated variously, more particularly with reference to the celestial sphere. This is an imaginary sphere of infinite radius, which is meaningful because we see astronomical bodies in the sky in projection on to this sphere and in spherical perspective. Thus if you look at the Sun and the Moon when both are in the sky, the fattest part of the lunar crescent does not face the Sun along a straight line, but along a great circle of the celestial sphere. Since the sphere is of infinite dimensions and parallel lines meet at infinity, any normal to a given plane will define the same point in the sky. Similarly the projection of any closed two-dimensional curve, such as an orbit, from any point inside it and within its plane will define a great circle on the celestial sphere. On the other hand, all inclinations are truthfully preserved in projection in the form of arcs, the full great circle containing 360° or 24 hours (Fig. 1).

This way of looking at things presents certain astronomical advantages. Thus, for instance, any normal to the plane of the ecliptic, which itself marks the apparent path of the Sun in the sky as well as the intersection of the plane of the Earth's orbit with the celestial sphere, will point at the pole of the ecliptic; any normal to the plane of the Moon's orbit will define its pole, and so on. Rephrased in these terms, Cassini's third law will read, 'the celestial poles of the ecliptic, of the lunar orbit and of the Moon lie on the same great circle and in this order.'

The first law calls for no special comment. The determination of the correct location of the lunar poles, however, is a ticklish observational task, owing to the slow rotation and the mountainous character of the polar terrain. The present figure of 1° 32′ for the tilt of the rotational axis of the Moon is due to K. Koziel, of

Cracow, who specializes in this type of work, and embodies an uncertainty of about 7″.[5]

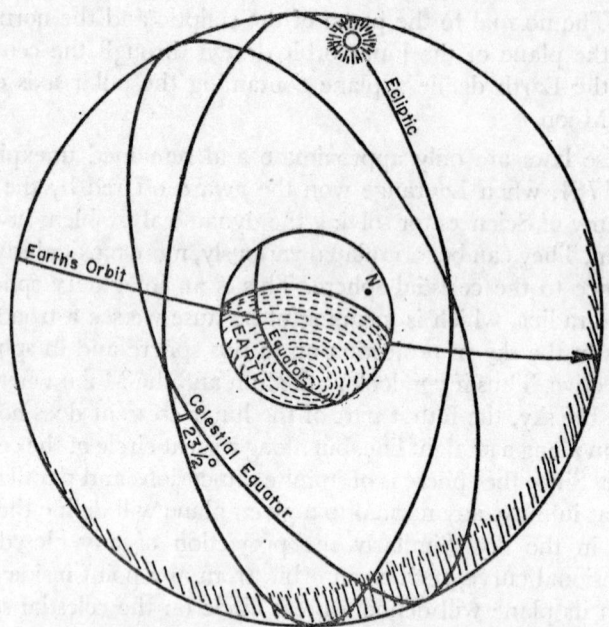

Fig. 1. The 'celestial sphere' which may be imagined as a kind of transparent plastic balloon with all the heavenly bodies projected on to it. The plane of the Earth's orbit intersects the sphere along the *ecliptic*, the apparent path of the Sun and the Earth's shadow. An image of the Sun has been drawn on it. Similarly, the plane of the Earth's equator intersects the sphere along the *celestial equator*.

A rotating body has gyroscopic stability, increasing with its angular momentum (p. 12), and tends to preserve a constant orientation with respect to remote objects, such as the fixed stars. This is held to be due to the attraction of the distant parts of the universe (Mach's principle) and underlies Cassini's second law. Secular changes apart, the axial tilt of the Earth is fixed relatively to the pole of the ecliptic, which – we know – images its orbit. But the lunar pole deviates from the pole of the lunar orbit by 6° 41′, and is true, instead, to the Earth's solar orbit, which may seem odd at first sight; but the reason is that the pole of the ecliptic is fixed among the stars, while that of the lunar orbit is not.

Orbital motion, too, enjoys gyroscopic stability, but if a persistent force is applied to the rotating body it is thrown into *precession*, in which its axis preserves a constant inclination relatively to a point fixed in the sky but describes a cone about it. In the case of the lunar orbit the constant inclination to the ecliptic is 5° 8′ 43·4″, and the period of precession (which in this case is negative – *recession*) is 18 2/3 years. During this period the *nodes*, or the points of intersection of the orbit with the plane of the ecliptic, *regress*, or move counter to the sense of rotation, round the latter. The disturbing force comes mainly from the attraction of the Earth on the Earthward bulge of the Moon (p. 12), which dips alternately north and south of the Earth-Moon line owing to the inclination of the lunar equator to the Moon's orbit.

This is not the only *periodic perturbation* of lunar motions; these are indeed so complicated that it is all too easy to get tied up in 'nodes'. The way to escape the confusion is by remembering their causes, which are not difficult to grasp.

While the line of the nodes regresses, that of *apsides*, or the major axis of the orbit, marked out by the perigee and the apogee, advances. The main cause of this is that the Moon moves not only round the Earth, but also and even mainly round the Sun. Being closer to it on the one side of the orbit than on the other, the Moon's circumsolar motion is now speeded up and now retarded relatively to the simultaneous progress of the Earth, approximately midway between. The effect is similar to that of the tides and makes the whole orbit turn round in its plane within 8·85 years. In fact, owing to the decrease in the gradient (steepness) of the solar field (see p. 12), the acceleration on the Sunward side of the orbit is slightly in excess of the deceleration on the opposite side, and the perturbation has a secular term as well, tending to blow out the lunar orbit towards Mars. Planetary perturbations come in further to complicate matters.

Another periodic perturbation is *evection*, which affects the eccentricity of the orbit and makes it fluctuate from 0·0432 to 0·0666. As the line of apsides revolves, it passes in and out of alignment with the direction towards the Sun and so with the line of *syzygies*, which joins the positions of the full and the new Moon. When the two lines coincide the pull of the Sun tends to elongate the orbital ellipse, and conversely when the two lines are at right angles to each other it tends to make it rounder. The period of evection is 31,807 days, and at its maximum evection may displace the Moon in longitude by $1° 16' 26·4''$.

There is also a *variation*, due to the fact that the gravitational attractions of the Earth and the Sun are aligned along the syzygies and at right angles to each other along the line joining the two quarter phases (half-Moon to half-Moon). The corresponding period is equal to half the synodic month, and the maximum longitudinal displacement, unknown to the Ancients and first discovered by the Arabic astronomer Abdul-Wefa in the tenth century, is $39' 29·9''$, or more than the Moon's apparent diameter. The eccentricity of the Earth's orbit leads to an *annual inequality* or $11' 8·9''$ in longitudinal amplitude, all of which affects the sidereal and synodic periods in the way alluded to on the earlier pages of this chapter.

But let us reconsider the recession of the nodes of the lunar orbit on the ecliptic and the line of syzygies. The ecliptic is so called because it is the great circle of the sky along which the Sun and, opposite it, the Earth's shadow travel and where eclipses occur.

A solar eclipse takes place when the Moon meets the Sun and a lunar one when it meets the Earth's shadow. This can happen only when the nodes come sufficiently close to the syzygies. A central eclipse, which may be total or annular if solar, requires an exact coincidence between a syzygy and a node. The Moon, however, is so close to the observer, and the Earth relatively so large that the line drawn from him to the centre of the Moon will form an angle with the ecliptic varying with his geographical position. In other words, the Moon will not be seen by all observers simultaneously at the same point of the sky. This is indeed how the lunar parallax, or the angle at which the Earth's equatorial radius is seen from the Moon, and hence the distance of the Moon and its true size, are conventionally obtained. But this rather complicates the geometry of eclipses.

The shadow cast by the Earth and the Moon consists of a long, gently tapering cone of full shadow, or *umbra* from within which no part of the Sun can be seen, and of a widening cone of partial shade, or *penumbra* within which the Sun is partly eclipsed. With the Moon at apogee and the Earth at perihelion (point of orbit closest to the Sun), the lunar umbra peters out short of the Earth and the solar ring in the annular eclipse attains its greatest width. Conversely, when the Moon is at perigee and the Earth at aphelion, assuming that the nodes oblige, the area of *totality*, where the Sun is completely covered up, will reach its maximum cross-section of 167 miles. The true stretch of totality in latitude and longitude will generally be greater owing to the oblique projection of the shadow cone on to the curved surface of the globe.

On the other hand, the visibility of a partial solar eclipse follows the opposite course, as the cross-section of the penumbra varies inversely to that of the umbra.

The mean diameter of the Earth's umbral cone at the Moon's distance is 9,202 km=5,705 miles, and so much larger than the diameter of the Moon (3,476 km). Hence it follows that total eclipses of the Moon must be relatively frequent; and, since the penumbral cone is wider still, penumbral eclipses, which will be seen from the Moon as partial eclipses of the Sun, will occur even more often. The average duration of a central eclipse is 3 hours 46 minutes, 1 hour 43 minutes of which will be total, but the eclipse will be longer with the Moon at perigee and the Earth at aphelion and *vice versa,* and in calculating the duration not only the sizes of the shadow, but the variations in orbital velocities must likewise be considered.

It is clear that lunar eclipses must be absolutely more numerous than the solar, but they are made even more so by the fact that an eclipse of the Moon can be seen from any point of the Earth's surface where the Moon is above the horizon at the time. A solar eclipse, on the other hand, is localized within that part of the Earth's surface which lies within the lunar shadow.

It is, further, clear that a solar eclipse can take place only at new Moon and a lunar eclipse only at full Moon. The *phase angle* is the angle defined by the lines drawn from the observer to the centre of the Moon and from the latter to the centre of the Sun. It is usually assumed for the sake of simplicity that the three points lie in the plane of the ecliptic. The phase angle is 0° at new Moon, waxes to 90° at the end of the first quarter of the lunation, at *dichotomy,* or 'halving', of the Moon, when the *terminator,* as the line dividing the day from the night hemisphere is called, is exactly in the middle of the disk. In the second quarter the phase angle is completed to 180° at full Moon, in the third to 270° at waning half-Moon, to return to 360° or 0° at the end of the fourth quarter.

This, however, is only a rough view of the situation. In reality the lunar orbit makes, perturbations apart, an angle of 5° 8′ 43·4″ with the plane of the ecliptic, which, moreover, is itself inclined at 23° 27′ to the Earth's equator, and the apparent position of the Moon in the sky is further affected by the geographical latitude and the season of the year. The Moon rides high in the night sky in winter and low in summer.

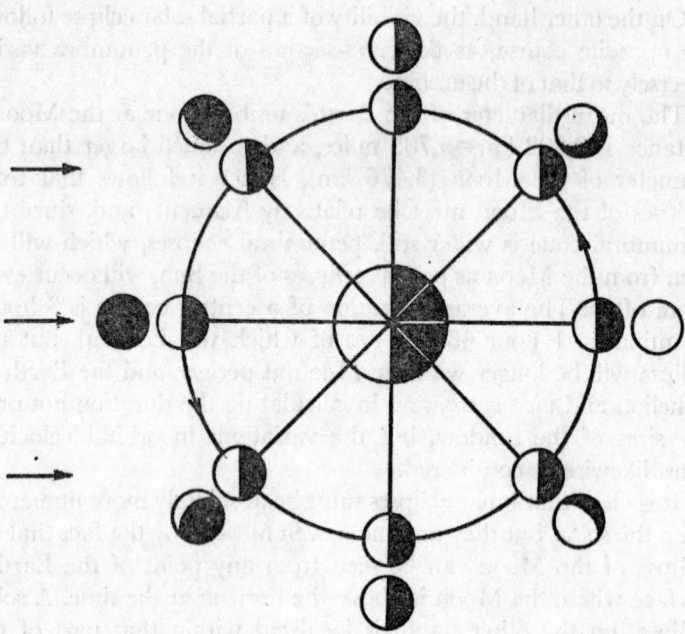

Fig. 2. Phases of the Moon. The Moon moves round the Earth counter clockwise as shown by arrows on orbit (circle). The arrows on the left show the direction of the sunshine. The angle between the line towards the Sun from the centre of the Earth and from it to the centre of the Moon is the *phase angle*; it is 0° for new Moon, 90° at first quarter (bottom), 180° at full, 270° at last quarter (top), and so on. The 'Moons' outside the circle show the phase as seen from the Earth when the Moon is in the corresponding orbital position.

Thus our view of phases is generally somewhat skewed, the line of maximum illumination and the line of sight being not in the same plane. We 'overlook' or 'underlook' the Moon by a small angle. This makes no difference at dichotomy, but at all other phases we usually see too much at one cusp and too little at the other. This effect, known as *phase defect* or *polar phase*, becomes particularly noticeable as the phase angle approaches the critical values of 0° and 180°. The new Moon will show a thin sliver of light at its north or south pole, depending on whether it is below or above the ecliptic and on how low or high it is in the sky. Conversely, a shadow will encroach on its opposite pole at full.

Thus a truly new Moon can be seen only during a solar eclipse, and a truly full Moon must inevitably be eclipsed by the Earth.

We have seen that an eclipse can occur only if the Moon is sufficiently close to one of the nodes and only at new or full Moon, which means that the line of the nodes must be passing through the Earth and the Sun. Since, though, this line revolves backwards in 18·61 years, such a situation will arise twice within this period, and twice within it the line of the nodes will stand at right angles to the Earth-Sun line and eclipses will be impossible. The conditions needed for an eclipse, however, are not satisfied unless the Moon is at the node at the right phase.

Phases are repeated after the synodic month of 29·53059 days, while successive passages of the Moon through a given node (*ascending node* where it rises above the ecliptic and *descending* where it passes below it) are divided by the *draconic month* of 27·21222 days. Therefore, any phase will recur at the same node after a lapse of time that is a common multiple of the two periods. It so happens that 223 synodic months are very nearly equal to 242 draconic months.

This period of 18 years 10¼ days (6,585·25 days) is known as *Saros* and was first discovered by the Chaldean priests in about 1,200 B.C., who used it to predict eclipses. An eclipse seen at a given place will recur after a Saros, but only with a certain licence, due to the various irregularities in the motions of the Moon, the simultaneous movement of the Earth round the Sun and the approximate character of the Saros itself. A total eclipse may be followed by a partial one, or *vice versa*.

Nowadays eclipses can be calculated with great precision for thousands of years back and forth, and all eclipses between 1,206 B.C. and A.D. 2,163 are described in some detail in Th. Oppolzer's *Canon der Finsternisse*, first published in 1887. The ancient eclipses have been used to date historic events associated with them, such as the battle of Arbel in 330 B.C., in which Alexander the Great defeated King Darius of Persia. The Chinese astronomers have kept regular records of eclipses since 1,136 B.C.

The immense complexity of the problem of lunar motions and appearances, which has been outlined in the preceding pages (for fuller mathematical treatment see Z. Kopal's *An introduction to the study of the Moon*[5]), has two opposite causes : the insecurity of the Earth's gravitational grip on the Moon due to its relatively great distance (see p. 21); and its proximity to the observer, resulting in parallactic shifts in the position and surface perspective of the lunar globe.

As we have seen (p. 26), the synchronization of the *true* axial period and the *mean* orbital period is almost perfect, except for the so-called physical librations, which are very small and will be discussed in the next chapter. If, therefore, the Moon rotated about an axis normal to the plane of its orbit and moved unperturbed in a circle in the plane of the Earth's equator an observer on the latter would see almost exactly one half of the lunar surface, except that, for one thing, the Moon being close, the tangential lines drawn from his eye to the lunar limb would fall somewhat short of the great circle bounding the Earthward hemisphere, and, for another, when the Moon was rising or setting he would look over the limb thus defined, now on the east now on the west side, by an angle of 57′ 2·6″, known as the *diurnal libration*.

It is as if the Moon wobbled from side to side like a balance thrown off its equilibrium. Indeed, the verb *librate* and the noun *libration*, as well as equi*librium*, are derived from the Latin *libra*=balance.

The diurnal libration is a parallactic effect, arising from the changing position of the observer as he is carried along by the rotation of the Earth, and it is *substantially* longitudinal. But the same effect can be reproduced in latitude if the observer travels sufficiently far north or south for the Moon to sink to the horizon at culmination (highest point of the apparent path in the sky). Once again he will be overlooking the limb, this time at the north or the south pole, by 57′ 2·6″.

A further geometrical libration is caused by the inclination of the lunar orbit and of the lunar equator to the ecliptic, which add up to a maximum tilt of the one or the other pole towards and away from the observer respectively equal to 6° 51′. In combination the two librations yield a latitudinal wobble of 7° 48′ 2·6″.

The main libration in longitude has a different origin. It amounts to 7° 57′ and is often described as *optical* or *geometrical*, but I think *orbital* would be a better term. It is due to the fact that the angular velocity of the Moon's spin is constant to all intents and purposes (p. 26), but its orbital speed varies with the changing distance from the Earth and to a smaller extent from the Sun. Thus it now outruns rotation, now lags behind it. We may here heave a sigh of relief, as for once the adjective *longitudinal* does not require qualifying. Nevertheless, the observer's position remains important and to get the full benefit of the libration he must be so placed as to have the plane of the Moon's equator in his line of sight, a condition not easy to satisfy.

In sum total only 41 p.c. of the lunar surface stays permanently invisible from the Earth and used to provide a theme for conjecture and a playground for fictional imagination. The gain to the eye, however, is rather precarious, as the limb regions are seen in extreme foreshortening, and, owing to the crowding of mountain-sides, their topography is difficult to disentangle.

Rectified photography (pp. 5 ff.), whereby the flat photographic image of the Moon is projected on to a sphere and can be viewed or photographed there, restored to the original spherical spacing, affords a partial solution of the problem. The device is nearly foolproof so long as all the features are at the same level on the Moon's surface. Thus the crater outlines and the distances between them can be accurately portrayed, although an increasing loss of light and contrast towards the limb is inevitable. Slopes are much more tricky. The camera, or the telescope, by means of which the 'flat view' is obtained, can never reach the slopes that face away from it, but it does see those turned towards it. A rectified photograph cannot show anything that is not in the 'flat' one – it can only make it stand out more clearly. As a result in a rectified photograph all craters and depressions appear steep and shallow on the nearer side and, contrariwise, gentle but deep at the farther end, this vertical distortion being intensified as the limb is approached.

This method was used to some advantage, but its triumphs were rather short-lived, as no sooner had it got under way than direct views of the limb regions and the averted hemisphere began to come in from orbital satellites.

Lunik 3 was launched on 7th October 1959 and heralded the new era by radioing back to Earth two days later the first television pictures of the other side of the Moon.[2] Their quality left much to be desired, and the map constructed from them after much labour has proved to be inaccurate in many important respects, but a beginning was made. In 1965 the Russian *Zonds* and especially the American *Orbiter* probes followed in quick succession, with greatly improved definition, and by the time of writing all the remaining gaps in the map of the averted hemisphere have been filled. It is as accurate and detailed as that of the more familiar side of the Moon, except for the scarcity of recognized names and height measurements.

We even have photographs of an eclipse of the Sun by the Earth taken from the Moon, as well as of the Earth itself, which put in its proper place as just another planet. This was still more dramatically demonstrated in December 1968, when the *Apollo 8* made the first manned flight round the Moon.

The position of the Moon is undergoing a reverse evolution : from being a heavenly sight it is becoming a world, stripped of romance, but perhaps more awe-inspiring in its immediacy, mysteriously different from the Earth and yet alike in its general features. Thus selenography and geography have come close together.

Before relinquishing the subject of the motions and changing appearances of the Moon, which is of enduring validity, it may be worth mentioning that the phase of the Earth as seen from the Moon is always complementary to that of the Moon as seen from the Earth : they add up to full.

Apart from being lit by the rays of the Sun, the side of the Moon facing the Earth is also in receipt of the sunshine reflected by the latter. Its amount varies with the Earth's phase, being greatest at new Moon when the Earth is full and vanishing at full Moon when it is new. It is also affected by the differential brightness of the continents and oceans presented by rotation and such factors as cloud cover and the spread of the winter snows. Taking the Earth's mean visual albedo as 0·39, a full Earth will supply on the average the light equivalent to 76·2 full Moons and strongly bluish as compared with the 'golden' moonlight, which looks greenish to the dark-adapted eye only because this is the colour of rod vision : any faint light looks green at night.

To the terrestrial observer the Earthshine on the Moon appears as *ashen light*, faintly marking out the night hemisphere and growing in intensity as the crescent narrows. It is readily visible on the 'young' waxing Moon when it is popularly described as 'the old Moon in the young Moon's arms'. As a matter of fact 'the young Moon' could be seen 'in the old Moon's arms' with equal ease, except that few people get up early enough to surprise it in the act. Public opinion is often unfairly biased.

REFERENCES

1 ALLEN, W. C. (1963). *Astrophysical Quantities*. 2nd edn. Athlone Press, London.
2 BARABASHOV, N. P. *et al.* (1961). *Atlas of the Other Side of the Moon*. Pergamon Press, London.
3 FIRSOFF, V. A. (1961). *Moon Atlas*. Hutchinson, London.
4 FIRSOFF, V. A. (1964). *Exploring the Planets*. Sidgwick & Jackson, London, and 1968, A. S. Barnes & Co., Cranbury, N.J.
5 KOPAL, ZDENEK (1966). *An introduction to the study of the Moon*. Reidel, Dordrecht, Holland.

4

Figure and internal constitution

The tidal bulge of the Moon has been referred to repeatedly in the two preceding chapters. In about 1860 P. A. Hansen in Denmark calculated from lunar motions that the Moon's barycentre was displaced 33 miles away from the Earth towards the invisible hemisphere, so that the figure of our satellite would resemble that of an egg, its pointed end facing the Earth. Air and water would thus tend to concentrate at the opposite, providentially invisible, blunt end, while all we could see was an airless desert at the top of an enormous mountain.

This pleasing idea had a short life and was soon debunked by Simon Newcomb. We have seen on page 15 that the atmospheric tide raised by the Moon is insignificant because our air ocean has no shores. The high tides in constricted waters arise from the banking-up of the lunar swell against the coastlines, and a small island amid an open ocean is substantially tideless. We may, therefore, follow the example of the Russian in the story of 'The Elephant and the Five Nations' and ask, does the tidal bulge of the Moon really exist?

The conception of the Moon being a ball of liquid of uniform density may appear a little improbable at first sight. As a matter of fact, however, large planetary masses, owing to the rheid creep of the rocks (p. 16) and the response to stress of their liquid core, behave in the long run not unlike such a ball of liquid and tend to assume a figure of hydrostatic equilibrium. In 1959 Sir Harold Jeffreys viewed the problem from this angle and considered the

Moon as a triaxial ellipsoid : one axis x pointing at the centre of the Earth from the centre of the Moon; the other y being ranged at right angles to x and to a z axis. perpendicular to both and lying in the plane of, but not in exact coincidence with, the polar axis of the lunar globe. If we now call the mean semi-diameter (radius) of the Moon a, and the semidiameters measured along x, y and z axes, a_x, a_y and a_z respectively, the result of his calculations was :[4, 9]

$$a_x - a = +38 \text{ metres}$$
$$a_y - a = -11 \text{ metres}$$
$$a_z - a = -25 \text{ metres}$$

Such small departures from the sphere would be entirely lost among the surface relief, which includes on the Earthward side extensive dark plains, called *maria*, as well as crater uplands, or *terrae*, involving differences of level of up to 10,000 metres (6 miles). Now, although altitudes of the mountain tops above the country at the foot can be measured from the shadows with a considerable degree of precision, absolute heights present an awkward problem, as there exists on the Moon no ready references level, such as is provided by our seas.

Instead of the sea level, measurements have to be referred to the *mean sphere*, which can be winkled out only by a laborious observational technique based on librations.

As the Moon wobbles from side to side the relative positions of the features on its face change. On a perfect sphere the distance of any point of the surface from the centre is the same, and the angular shift in its apparent position due to libration is a simple geometrical function of its distance from the centre of the disk and from what may be described as the 'equator of libration' (that great circle of the sphere along which the shift due to libration attains its maximum value). Since, however, the actual distance of the various points of the surface from the centre of the globe varies, the observed angular shift will exceed or fall short of the theoretical value that corresponds to the mean sphere in proportion to the elevation or depression of the point relatively to this assumed level of reference. In this way the true shape of the Moon can be discovered and a *hypsometric*, or 'iso-level', map of the Earthward hemisphere can be compiled if we have a sufficient number of accurate measurements of angular shifts of selected points on the lunar surface.

The procedure is, therefore, to establish a reference network of small, readily identifiable craters and to measure the varying angular separation of these second-order points from one or more first-order points near the centre of the disk in dependence on libration. Even the libration itself can be obtained by this method. The exactitude of the individual measurements may fall a good deal short of the requirements, owing to the limited resolution of the instrument used, atmospheric unsteadiness (bad 'seeing'), irradiation, and purely accidental errors of judgement. In the long run, however, all these errors will be scattered impartially above and below the true value and tend to cancel out statistically more and more as the number of observations increases. Thus the true value can be found with increasing approximation by the method of least squares, so long as there are no systematic errors, to cause a biased deviation. All that is needed is time and infinite patience.

Observations of this kind have been carried on for quite some time, the small bright crater Mösting A in the region of Sinus Medii being usually taken as the first-order point.

In the earlier work, initiated by Bessel at Königsberg, in 1839, the measurements were made by means of a *heliometer*, so called because it was originally designed for determining the diameter of the Sun. A heliometer is a refractor with a split objective which is rotatable about the optic axis and whose one half can be slid sideways relatively to the other by means of a micrometric screw. When the two halves of the object glass are out of coincidence they produce two images of the Moon in imperfect overlap. To measure the separation between two craters, the objective is rotated so as to bring the craters on to the line dividing the halves, which are then moved until the images of the two craters fall together. Knowing the focal length of the instrument, the angular separation of the craters can then be obtained from the reading of the micrometric screw. There will be two such points of coincidence : one when the movable half is shifted to the right and the other when it is displaced to the left. This gives an equalizing check on the measurement, reducing the likelihood of bias.[1, 3, 5]

The apertures of the heliometers ranged from 5 to 7 inches, so that their resolution and, consequently, the accuracy of the measurements was not very high. Today they are considered obsolete. Only one heliometer, set up in 1891 at the Engelhardt Observatory, in Kazan, Russia, is still in use.

Elsewhere the heliometric method has been superseded by measurements made on photographs, taken with large telescopes of high resolving power. Nevertheless, the results are not very convincing. The hypsometric maps of the Moon drawn by different investigators do not seem to have much in common, that constructed by the Aeronautical Chart and Information Center (A.C.I.C.) of the U.S. Air Force apparently being the most reliable (Kopal),[4] inasmuch as it is based on a larger number of measurements and a unified procedure, but in view of the general lack of agreement it, too, should be taken with a pinch of salt.

Nevertheless, there is a consensus between the various workers that a tidal bulge of between 1 and 4 km, and most probably about 2 km, actually exists. R. B. Baldwin (1958) made separate estimates for the terræ and the maria and obtained the respective values of the bulge equal to $2\cdot26\pm0\cdot21$ km (probable error) and $2\cdot03\pm0\cdot37$ km. Thus the bulge would be present in both types of surface feature and be of the same order, although the uplands average $1\cdot74$ km (5,710 feet) higher than the maria.[9]

The determination of the polar axis similarly presents great difficulties, aggravated by the existence of high mountains (the Leibnitzes and the Doerfels) near the south pole. Baldwin obtained a polar semidiameter a_z that is somewhat longer than the east-west semidiameter a_y, though shorter than the Earthward semidiameter a_x. Limb-to-limb measurements, however, indicate that the shortest diameter of the lunar globe coincides with neither and deviates by an angle of $55°\pm2°$ east (conventional) from the south pole, that measured $35°$ west of the south pole being $2-3$ km longer.[4]

The existence of lowlands and uplands would by itself lead one to expect something of this kind, but the issue is somewhat complex and closely linked up with the problem of the moments of inertia of the lunar globe.

The *moment of inertia* measures the resistance of a mass to a change of its state of rotation or oscillation (reciprocal rotation), which may be zero, about a given axis. It involves two terms:

the total mass and its distribution relative to, or distance from, the axis of revolution.

The Moon revolves about its polar axis and is distorted by the tidal pull of the Earth. If, however, it were in hydrostatic equilibrium, the resulting stresses would nowhere cause it to deviate from the true sphere by more than 40 metres (p. 39). Not only would such inequalities be unobservable, but they could not observably affect the motions of the Moon. Cassini's third law, however, demands an appreciable action of the Earth on a lunar bulge to cause the regression of the nodes (p. 29).

The moments of inertia of the lunar globe about the axes x, y, z, where z is taken to coincide with the polar axis, are usually called A, B, C.

Cassini's relationship requires (Jeffreys, 1961) the existence of a bulge along the axis x (towards the Earth) equal to 1·1 km on the assumption that the Moon is a homogeneous body. This corresponds to the *dynamical ellipticity* $\beta = (C - A)/B = 0\cdot0006279$. If the material within the Moon varies in density, radially, circumferentially, or both, the geometrical bulge may be larger or smaller, so long as it results in a distribution of mass that satisfies this relationship. The bulge defined on the assumption of homogeneity is called the *dynamical bulge*.[3]

The librations bring the dynamical bulge out of alignment with the Earth, which tries to pull it back into it against the moments of inertia along $x, y,$ and z. This results in *physical librations*. These additional wobbles do not exceed 2′ in amplitude as referred to the Moon's centre, which corresponds to 0·54″ at the centre of the lunar disk as seen from the Earth. In the case of most observations this falls below the limit of telescopic resolution. The determination of an angle below this limit is possible in principle, being given a sufficiently long series of observations free from systematic errors. The issue is similar to, but somewhat different from, that of determining the geometrical shape of the Moon from the librational shifts of selected points of the surface relatively to a point or points of reference: what is measured in this case is the deviation of the observed shifts from their values expected from the known geometrical librations. The same sets of observations can be used for both purposes.

The Herculean labour of reducing the vast observational material, both heliometric and photographic, to distil out of it the physical librations and so A, B and C, has recently been carried out by Koziel at Cracow and resulted in the following dynamical ellipticities[5] :

$$\alpha = (C - B)/A = 0.0003984$$
$$\beta = (C - A)/B = 0.0006294$$
$$\gamma = (B - A)/C = 0.0002310,$$

whence the *mechanical ellipticity* $f = \alpha/\beta = 0.633$, and the inclination of the polar axis $I = 1° \ 32' \ 4''$, as quoted on p. 28. It will be noticed that Koziel's value of β is slightly higher than Jeffrey's theoretical determination, but otherwise the agreement is good.

And you may well ask, 'So what?'

In the first place, it is clear that the Moon is not in the condition of hydrostatic equilibrium; in other words, it does not behave like a homogeneous liquid. In the second, the uneven distribution of mass within the globe will result in stresses, which must be compensated by some means or other.

In 1908 G. H. Darwin calculated that if the Moon were of uniform density a bulge of 1 km would create at its centre a stress of 19 bar = 19,000 kg/cm[2]. This led H. C. Urey to suggest (1952) that the Moon had never been molten, had no liquid core, and possessed 'the rigidity of brick'.[10] He later came to modify his ideas, and jointly with Elsasser and Rochester proposed in 1959 a hypothesis that local differences in surface density had arisen from the infall of meteoritic matter. As S. K. Runcorn points out, this requires meteoritic material to have been systematically less dense on the side facing the Earth.[9] On the other hand, the proposition that the Earth could have shielded this side against meteoric bombardment, so that more meteorites struck the other side of the Moon, contributing more denser matter to its outer layers, appears plausible, except that it would let the biggest fish escape; for if the marial basins were marks of meteoritic impact the Moon would have sustained the mightiest blows on its Earthward half.

Jeffreys attempted (1924) to account for the Moon's departure from the figure of hydrostatic equilibrium (p. 38) by assuming that it had solidified at an early stage on tidal evolution when it orbited the Earth much closer, but already was in synchronous rotation. The bulge would thus be 'fossil'. It would, nevertheless, have to satisfy the hydrostatic requirements and yield a mechanical ellipticity of either 0·25 or 0·75, whereas its actual value is 0·663.

Logically unexceptionable, however, as such arguments may be, they are largely robbed of practical interest by the simplifying assumptions made at the outset and the either-or attitude. Thus the Moon is usually thought of as an essentially homogeneous body, which in the light of our knowledge of the internal constitution of the Earth and of the composition of meteorites appears highly improbable. The change in the position of the polar axis may have occurred, either through the slipping of the crust over a yielding substratum, or the lunar globe may have toppled over in the course of tidal evolution, as was suggested by G. P. Kuiper in 1954.[6] The simultaneous changes in the inclination of the orbit (see pp. 12 ff.) must also be taken into consideration. Yet the most important point is that there can be no intrinsic reason why the present situation must be due to any single cause : this is a matter of verbal neatness, not of physical necessity.

Thus in 1964 B. J. Levin in the U.S.S.R. proposed to account for the lunar bulge by a secular thermal expansion of the equatorial zone, due to differential insulation. The proposition was analysed mathematically by Z. Kopal,[4] who found that, although an effect of this kind should be present, it could not yield a dynamical ellipticity of over 0·00005, which is at least an order of magnitude too small. Levin, however, defends his position in a later paper and claims an excess of the equatorial over the polar semidiameter equal to 1,250 metres, and so even larger than its 'observed' value.[7] All this is based once more on the assumption that the Moon is a homogeneous body; and the extremely low thermal conductivity of its surface formations makes the whole argument highly suspect.

On the other hand, Runcorn argues (1962)[9] that the differential distribution of mass within the Moon can be accounted for by a four-cell thermal convection system in the semi-molten mantle with a slow creep of about 1 cm per year. Here the Moon is no longer homogeneous and has a layered structure resembling that of the Earth, including a small liquid core. G. Fielder finds (1965)[1] that Runcorn's theory yields a stress pattern closely comparable to that inferred from the *tectonic grids* of the Moon, which will be considered more fully in another connection. There would thus be some observational evidence as well as a geological analogy in support of internal flow.

Once more the conception has been criticized by Kopal.[4] He contends, on the one hand, that the surface relief observed at the limb involves differences of level of 8 – 9 km, which, jointly with the deviations of ± 2 km from the mean sphere over large areas of the lunar globe, preclude a yielding interior, so that the Moon could never have passed through a molten stage. On the other, he maintains that the rate of flow would be some 10 – 8 cm/sec and so insufficient to establish a steady convection pattern in 10^9 years.

Runcorn's rebuttal (1967)[9] centres on the point that Goudas's calculations, on which Kopal's latter argument is based, overlook the distinction between the maria and terrae, introduced by Baldwin. Once this is taken into account, the alleged discrepancies (fourth harmonic) disappear.

It is not easy to give an exhaustive picture of so difficult and specialized a field of study in the space of a few pages. But in regarding it as a whole one is left with the strong impression of an elaborate mathematical game in which parameters have replaced the chessmen. The game may be exciting; it certainly generates strong feelings in the players. Nor can it be denied that it has some bearing on the real situation, except that this bearing is so vague, the conclusions so uncertain. There is the real 'honest-to-goodness' Moon up in the sky – almost underfoot, and it seems to be having a good laugh at this abstruse scholastic pastime.

The equations may decree a rigid interior, but the real Moon shows undeniable slip and wrench faults, some of the latter involving horizontal displacements of many miles, overthrusts, subsidences and upheavals, which all bear witness to crustal movements over a yielding substratum.[1, 2]

It may fit nicely some theoretical conception that the Moon is a cold, undifferentiated spheroid of an uncompressed mean density 3·27, which rises to 3·40 at the centre under the pressure of the overburden. This may correspond to the density of dunite, a heavy ultrabasic rock suspected (not known) of forming the mantle of the Earth, or of similar material found in chondritic meteorites, but having a lower content of radioactive elements, to prevent internal melting and the consequent fractionation according to specific weight. But the real Moon cares not a whit about all that fine logic : its surface displays undoubted flows of lava and other familiar marks of igneous activity, and so of internal heat.[1, 2, 8] And is this logic all that fine? To sustain such a position, one must at least suggest some credible process of chemical segregation that could have allowed such a peculiar body to arise in the first place. Manipulating parameters is not enough.

To sum up, it does not look as if our understanding of the condition and structure of the Moon will be greatly advanced by concentrating on lunar motions and bulges.

We must try a different approach.

REFERENCES

1 FIELDER, GILBERT (1965). *Lunar Geology*. Lutterworth Press, London.

2 FIRSOFF, V. A. (1961). *Surface of the Moon*. Hutchinson, London.

3 KOPAL, ZDENEK (1966). *An Introduction to the Study of the Moon*. Reidel, Dordrecht, Holland.

4 KOPAL, ZDENEK (1967). 'The shape of the Moon, its internal structure and moments of inertia', *Proc. R.S., Series A,* **296,** No. 1446, pp. 254–265.

5 KOZIEL, K. (1967). 'Differences in the Moon's moments of inertia', *Proc. R.S., Series A,* **296,** No. 1446, pp. 248–253.

6 KUIPER, G. P. (1954). 'On the origin of lunar surface features', *P.N.A.S.U.S.A.,* Vol. 40, No. 12, pp. 1,069–1,112.

7 LEVIN, B. J. (1967). 'Thermal effects on the figure of the Moon'. *Proc. R.S., Series A,* **296,** No. 1446, pp. 266–269.

8 MOORE, PATRICK and CATTERMOLE, PETER (1967). *The Craters of the Moon*. Lutterworth Press, London.

9 RUNCORN, S. K. (1967). 'Convection in the Moon and the existence of a lunar core', *Proc. R.S., Series A,* **296,** No. 1446, pp. 270–292.

10 UREY, H. C. (1952). *The Planets*. Oxford University Press.

Abbreviations used

Proc. R.S. = Proceedings of the Royal Society.

P.N.A.S.U.S.A. = Proceedings of the National Academy of Sciences of the U.S.A.

5

The Moon as an Earth-like body

The disparity of the Moon and the Earth in size and mass is considerable, but not overwhelming, and, as stated in Chapter 2, they naturally combine to a double planet rather than a planet with a normal satellite. To recapitulate, if we take the Earth data as the respective unit of comparison, those for the Moon are :

Diameter	Volume	Mass	Mean Density	Surface Gravity	Velocity of Escape
0·2725	0·0203	0·0123	0·7626 or 3·342 gm/cm³	0·165 or 162 cm/sec²	0·213 or 2·38 km/sec

The last figure in the table is 11·18 for the Earth. The mean density of the Earth works out at 5·517 gm/cm³. The two bodies could not have originated in mutually remote parts of the primordial cloud that gave rise to the Solar System and so must have been of closely comparable chemical composition at birth.

They are assumed to have been born cold from small solid masses of lithophilic (rock-forming) matter, either metallic or stony, varying in dimensions from dust grains to sizeable meteorites, and 'ices', trapping between them a certain amount of gas. These materials are thought to have resembled present-day comets. The meteorites, however, which are sparsely pelting the Earth to this day are certainly of secondary origin : they show signs of high pressure and great heat of formation, and it is a moot question whether their parent bodies have been of planetoidal dimensions or they have come from a larger planet or planets that once orbited the Sun between Mars and Jupiter and were destroyed by a collision or some other catastrophe. In any event their original counterparts may be expected to have been very lightly consolidated under conditions of near-zero gravity, possibly collecting into cloud-like swarms, still represented by the somewhat mysterious objects known as 'nebulous meteorites'.

The ices would have been frozen water, ammonia and hydrocarbons, although it has been suggested that the temperatures within the orbit of Mars at the time of aggregation of planetary masses were too high for ices to form or persist, in which case these substances and other vapours or gases could have been retained by sorption, solution and chemical bond (hydration, ammoniation, etc.).[19] As the photoplanetary mass grew gravitation alone would become sufficient to bind gases, but the almost total absence of the cosmically abundant noble gases from the atmosphere of the Earth goes to show that the primary gas envelope round it has been completely lost during the ensuing hot stage.

The contraction of the originally diffuse or lightly compacted protoplanet would generate heat and increase its rate of spin. The decay of radioactive matter, originally much more abundant than now (p. 19), would provide a further and even more potent source of internal heat. The initial contribution would come mainly from the heavy radioactive isotopes with short half-lives, such as I^{129}, Pd^{107} and Al^{26}, while the liberation of energy by the long-lived species, K^{40}, Th^{232}, U^{235} and U^{238}, would continue at a lower rate to the present day.

This problem has been investigated by Urey (1952), MacDonald (1960) and Kopal (1962), who has shown that the Moon would lose its initial heat in a time short as compared with its age.[13] But the important point is that, as Kuiper has shown (p. 20), all bodies of and over 100 km in diameter must have experienced internal melting, and there can be no reasonable doubt that wholesale melting must have occurred in the Moon.[14] Thus iron, nickel and other heavy metals, which have melting points below those of the silicates, would tend to flow inwards and collect at the centre, while lighter matter would rise towards the surface. Some of the volatiles would be driven off. The increasing radiation of the Sun would ionize and dissipate the surrounding atmosphere, some of which could even have been shed by the centrifugal force of accelerating spin. As already mentioned, however, we can leave the primary atmosphere completely out of consideration.

A parallel concentration of the heavy radioactive elements, uranium and thorium, in the crustal and/or subcrustal layers must also be expected, as these large atoms become preferentially fitted into the looser lattices of the lighter acid rocks. This effect is well known from the Earth's crust.

The result would be a layered structure comparable to that of the Earth, as revealed by the study of seismic waves.

The uppermost layer or *crust* of the Earth consists predominantly of acid (over 66 p.c. of SiO_2) rocks, typified by granite, and the products of their decay, jointly known as *sial* after silicon and aluminium as their chief lithophilic elements. Sial has a mean density of about 2·70 and an average continental thickness of 35 km, being largely absent from the deep-ocean floors. It is bounded downwards by the *Moho discontinuity* of variable depth, where the mean density of the rocks jumps to 3·32. Below the Moho silicates of magnesium form the main rock-building minerals, whence this part of the globe is called *sima*. Sima composes the 2,900 km-thick *mantle,* which behaves as solid in transmitting transverse seismic waves, fractures under sudden stress, as deep-seated earthquakes show, but yields to gradual pressure, which enables it to sink and rise under differential crustal loading (isostasy, see p. 16) and from slow currents, shifting continents in the course of geological ages.

The temperature and so the rate of flow of sima increases downwards, the mean density rising simultaneously to 5·68, it is thought, wholly or chiefly owing to gravitational compression and changes of phase (crystalline habit), at a further discontinuity that divides it from the liquid *core*. Here a sudden increase in density to 9·43 is attributed to a change in chemical composition. By analogy with the nickel-iron meteorites the core is assumed to consist predominantly of these metals and is referred to as *nife,* accordingly. It is 3,430 km deep, its central portion behaving once more like a solid and being probably a highly compressed gas at a doubtfully estimated central temperature of $+14,000°$ C.[1] The central density of 17·2 is usually attributed to compression alone, although an increasing proportion of heavy metals may contribute to the effect. The main reason for discounting the latter lies in their low cosmic abundances.

Contrariwise, iron is second only to magnesium among the metals in the order of cosmic abundance and is represented handsomely among the meteorites, which may be taken as a fair sample of original lithophilic materials.

The low mass of the Moon could have affected its ability to retain gases, but it could not have prevented it from capturing proportions of heavy atoms, including iron, comparable to those found in the Earth. The Moon must have a nife core.

Tektites may or may not have come from the Moon, but at the time of writing we already have the first rough estimates on the composition of the lunar surface, based on the scattering by it of alpha particles and protons emitted by the *Surveyors V, VI* and *VII*.[20, 22] The first of these landed in a small crater in the southern part of Mare Tranquillitatis, the second on flat ground in Sinus Medii, and the third in a highland region, on the outer slopes of the crater Tycho.

The analysis is based on the Bragg rule, whereby the mass stopping power for alpha particles is inversely proportional to the square root of the atomic weight. The alpha particle, being a helium nucleus of atomic weight 4, is a comparatively heavy and insensitive 'missile'; the proton of atomic weight 1 is more discriminating for lighter atoms. In sum total the analyses summarized below begin with carbon and exclude elements lighter than beryllium, which may be important, as such elements form soluble compounds that are normally leached out of our rocks, but in the absence of surface water may be able to stay on the lunar surface. Further 'Ca' refers to elements with atomic weights between about 30 and 47, including, for instance, P, S, K, as well as Ca. 'Fe' denotes elements having atomic weights between approx. 47 and 65, and so Cr, Ce and Ni among others in addition to iron. In the case of *Surveyor V* the two groups are lumped together. The lower limit for Fe within the latter bracket has been set at 3 p.c.

With these reservations the table reads :

Element   Atomic Composition in p.c.

	Surveyor V	*Surveyor VI*	*Surveyor VII*
C	<3	<2	<2
O	58±5	57±5	58±5
Na	<2	<2	<3
Mg	3±3	3±3	4±3
Al	6·5±2	6·5±2	8±3
Si	18·5±3	22±4	18±4
'Ca'	} 13±3	6±2	6±2
'Fe'		5±2	2±1

The corresponding proportions in the igneous rocks of the terrestrial crust are :

O, 46·6; Si, 27·72; S, 0·052; Fe, 5·0; Co, 0·0023; Ni, 0·0080; Al, 8·13; Na, 2·83; C, 0·032; Mg, 2·09.

There are some differences. Thus there seems to be somewhat less silica on the Moon and more metallic oxides or else water of hydration, but the general resemblance is striking. The first two lunar samples may be taken to refer to the dark-coloured *lunabase* and the third to the light-coloured *lunarite*, and the differences between them support the general surmise that the latter is more acidic than the former. Detailed comparisons show the Sinus Medii site fitting a composition intermediate between basaltic achondrite and granite.[21] The lunarite, on the other hand, would be more acid than diorite and less so than granite,[22] with a higher proportion of felspathic minerals; in fact, rather like syenite.

The precision of such estimates should not be exaggerated, but they warrant two conclusions: the mean proportion of iron at least in the maria is comparable to that of our crustal rocks, and some vertical fractionation of material according to atomic weight has occurred on the Moon, just as it has on the Earth.

This raises an awkward problem. The sialic layer of the Moon has been put tentatively by Hédervári,[10] on the assumption that the Moon has no core, at 28·5 km (say, 17·7 miles), which is of no great consequence for its mean density. Not so the nife core. The high densities in the Earth's interior are due mainly to gravitational compression, which is unimportant in a body of lunar mass, so that it may be neglected in the first approximation. Even so the uncondensed density of nife will be about 8.

Among meteorites 'stones' are approximately 10 times as numerous as 'irons', though they, too, often contain a high proportion of ferrous material. Let us, therefore, assume for the sake of argument that the Moon's mass includes only 10 p.c. of ferro-nickel – and this is very much less than does that of the Earth. In order to yield a mean density of 3·34, the remaining 90 p.c. of the lunar mass must average 2·82, which is not much above the mean for sial, and both the composition of the Earth and of the stony meteorites indicates a great preponderance of heavy basic rocks. We must somehow accommodate a substantial mantle of sima, whose uncompressed density is estimated at 3·27. This results in a great deficiency of mass in the uppermost layers of the lunar globe.

Comparison with other Moon-size bodies is instructive in this context.

Neptune's Triton is credited with a diameter of 3,700 km (Moon, 3,476 km) and a density of 5.1, which would make it intrinsically heavier than the material of the Earth, but is doubtful in view of the difficulty of accurately determining the diameter of relatively so small and remote a body. The same applies to the satellites of Uranus, which are much smaller than the Moon, but are assigned densities from 4 to 6. In the systems of Saturn and Jupiter our footing is surer. Io, with a diameter of 3,200 km, has a mean density of 4·1; the still smaller Europa (2,900 km) at 3·7 is also denser than the Moon. All the other large satellites are less dense: Ganymede (5,000 km) – 2·4; Callisto (4,500 km) – 2·0; Titan (4,800 km) – 2·3.[9]

The relatively high densities of Io and Europa indicate that the effect cannot be ascribed to the scarcity of lithophilic materials, which would presumably include radioactive elements in the usual proportions. In any case Callisto has been found to radiate more heat than the Sun can provide, and the excess must be radiogenic.

It is usual to invoke the presence of large amounts of water and/or ammonia ice as rock minerals in the subsurface layers of these bodies, to account for their low mean densities. This is probably true, but we have just seen that the internal heat ought to be sufficient to melt this ice, though the escape of the liquid to the surface may be prevented by the permafrost seal.[4, 5]

Although the composition of the Jovian satellites may depart considerably from that of the Earth and the Moon, this lesson is applicable to the latter. The Moon, too, may be expected to contain large amounts of water (if not of ammonia) in its interior. This water would freeze upon reaching the permafrost layer below the surface, where massive accumulations of ice may form.

This would go some way to redress the balance. But by itself it is not enough, all the more so as the Moon is supposed to have once rotated in a matter of hours (pp. 18 ff.) rather than days, which would have prevented permafrost conditions, except in the polar regions. The Earth and the Moon must have followed different courses of development chartered by the disparity of their masses and surface gravities.

J. E. Spurr[18] investigated the compressional distortion of the old lunar ring mountains and concluded that since the time which he calls 'the Imbrian revolution', when the basin of Mare Imbrium and most other maria were initiated, the equatorial diameter of the Moon had shrunk by between 250 and 300 miles and the polar diameter by not less than 200 miles. These changes may have been occasioned by, or associated with, the hydrostatic adjustment of the Moon's figure to the slowing down of the speed of rotation, leading to the drawing-in of the equatorial belt and partial uplift of the previously depressed polar regions. The 'wandering' of the polar axis and the formation of the tidal bulge may also have to be considered. Such causes, however, could not by themselves have led to an overall shrinkage of the Moon in volume.

Spurr concludes in *The Shrunken Moon,* p. 8: 'Part of the equatorial shrinkage, and all of the axial shrinkage was general, attendant on, and no doubt due to, the steady enormous gas-exhalation. This conception is startling, for it indicates evolution of the present Moon from a more gaseous condition and larger bulk, by the loss of primitive gases.'

He expects the original diameter of the Moon to have been over 3,000 miles as against the present figure of 2,160 miles. This estimate is probably exaggerated, as Spurr starts from the assumption that the old ring mountains were originally circular and became compressed into hexagonal outlines, which may be questioned on the grounds of the general conformity of the ring walls with the lie of the land, more particularly as expressed in the grid system (p. 5). Nevertheless, there can be little doubt that some compression has occurred. In particular there is good evidence for equatorial shrinkage, resulting in a lateral compression of the crater shapes, which Spurr calls 'appression' and the breaching of the north and south walls.

To sum up, it seems that the original density of the Moon was much lower, and the Moon resembled the present condition of Callisto, which it might have retained, except for its capture by the Earth and the disturbance of the internal equilibrium caused by it. Thus on top of having to account for the present outward defect in the mass of the Moon we must also make provisions for a still greater defect in the past.

Of further interest in this connection is the analysis by N. F. Ness *et al.*[16] of the magnetic measurements by the lunar satellite *Explorer 35*, placed in a selenocentric orbit on 22nd July 1967 with a *periselene* (point closest to the Moon) of 800 km from the surface.

No disturbance in the magnetic wake of the Earth has been detected when the Moon was passing through it, whence it is inferred that, interpreted as a magnetic dipole, the Moon cannot have a magnetic field of its own in excess of 4×10^{20} cgs units or 8γ, although it could have local magnetic fields on the surface in a multipole system. This, however, disagrees with the Russian measurements made with the *Luna 2* and *10*, the latter of which had a closer orbit, with an aposelene at 2,700 km and periselene at 2,100 km or about 360 km from the surface. The *Luna 10* registered a magnetic field of $24 - 40\gamma$ regardless of its position relatively to the Moon and the Earth, on the basis of which Sh. Sh. Dolginov *et al.* (1967) conclude that the Moon has a magneto-sphere, as one might otherwise expect from its effect on the ionized layers in the Earth's atmosphere (p. 15).

If, however, we accept the *Explorer* data the Moon does not generate a magnetic shock wave in the solar wind, which, there-fore, may – it is suggested – be reaching its surface unhindered. This state of affairs necessitates a very low average effective con-ductivity of lunar material, not exceeding 10^{-5} mho/metre. With the conventional model of the Moon such a low conductivity cannot be reconciled with the high internal temperatures that must result from a uniform composition resembling chondritic meteorites. Indeed, Ness *et al.* put the mean temperature of the Moon at not over 1,000° K, which is still compatible with higher temperatures at the surface, where heavy radioactive matter would be preferentially concentrated in acid rocks, and a small liquid core. On the other hand, it is stated that the conductivity of 'impure water' would be too high to satisfy the equations, no reason, though, being given why this water should be muddy, which suggests a disturbed condition.

These results, even if confirmed, shed no dazzling new light on the internal constitution of our satellite. But they add to the embarrassment of the conventional point of view, and it deserves notice that low electric conductivity goes with low thermal conductivity, inferred from the temperature changes observed during lunar eclipses and from the radiation in millimetre and centimetre wavelengths, which in turn is associated with the porous or bubbly structure of the surface rocks that may be considered firmly established.

To recapitulate, neither the general principles nor what is already known about the chemistry of the lunar surface permit us to fall back on a composition drastically different from the terrestrial to explain the anomalous behaviour of the Moon. This extends to the proportion of radioactive matter, responsible for the internal heat. The answer must be sought in parallel but different courses of development of the Moon and the Earth.

Past histories worked out from a single state[13] are not very useful. A development is a continuous process, or rather an interplay of various processes going on simultaneously, and this is what makes a rigorous numerical approach very difficult. In comparing the probable evolution of the two planets from their days of birth as individualized masses in the Solar System, we may profitably bear in mind the following points :

1. The degree of consolidation of primordial material within the Moon will have been much lower from the start, especially on the periphery.

2. The evolution of heat, whether compressive or radiogenic, will have progressed from the centre outwards.

3. The ratio of the surface area to the volume of the sphere increases in inverse proportion to its radius. Thus the surface area of the proto-Moon will have been 10 – 15 times larger in relation to its volume than that of the proto-Earth. Other things being equal, the generation of radioactive heat will be proportional to the volume, while its loss by radiation will be proportional to the

surface area. The surface of the Moon will thus have been cooling 10 – 15 times faster. The issue however is complicated by the low thermal conductivity of the outer unconsolidated layers of the lunar globe, the result of which should be a comparative increase in internal heat paralleled by a lowering of the surface temperature.

The compressive stress at the centre of the Moon is barely comparable to that of the Earth's upper mantle. Owing to this the initial heating will be much more central in a body of the Moon's mass, and low conductivity will delay its outward progress. It is, therefore, in the central region of the Moon that melting will ensue, with a reshuffle of material according to specific gravity, the heavy metal phase sinking towards the centre and the liberated gases striving to escape upwards.[17]

Under high-gravity conditions the volatiles will be squeezed out to the surface much more effectively and be subsequently lost during the hot-surface stage of development (p. 48). But under low gravity gravitational fractionation will generally be less effective. The ascending volatiles will be chilled in approaching the surface layers, with at least partial condensation of such substances as water. These layers, being both light and cold, will yield rather than break under the pressure of the upbubbling gases. Thus the building of a metallic core and surrounding mantle, which must be accompanied by consolidation and so a decrease in volume, will be paralleled by an opposite process of inflation by the gases occupying the resulting gap between the hot interior and the still-cold outer layers. As the temperature continues to rise, the consolidated inner portion will advance, but it will also swallow the surrounding uncompacted matter, reducing its volume, while the volatiles will continue to expand upwards, insinuating themselves into any weaknesses they may encounter. The result will be a honeycomb of gigantic drusy cavities, filled with gas and water at a steadily rising temperature and pressure. There are, indeed, indications that similar gas bubbles exist in the mantle of the Earth, where they betray themselves as gravity anomalies, and may be responsible for some of the deep-seated earthquakes.

If this process continues indefinitely a point must be reached where the gases break through to the surface, causing a widespread volcanic activity, and form an atmosphere or be lost to space by

Plate 6a. Steep walls and huge masses on its bottom appear in this close-up photograph of a crater on the far side of the moon. Taken by the Lunar Landing Vehicle of *Apollo 10.*

Plate 6b. View from the right hand window of the *Apollo 11* Lunar Module showing the surface of the moon in the vicinity of the touch down. Numerous small rocks and craters can be seen between the module and the lunar horizon.

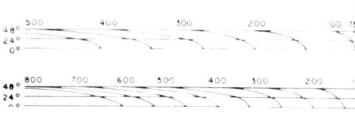

80° 90° 100° 110° 20° 130° 140° 150° 160° 170°

48°

40°

GAUSS

RAYLEIGH

30° JOLIOT CURIE LOMONOSOV

HUBBLE

MARE MOSCOVIENSE

20°

MARE MARGINIS

10°

0° MARE SMYTHII

10°

20°

TSIOLKOVSKY

HECATAEUS

HUMBOLDT

30° BARNARD

ABEL

MARE AUSTRALE JULES VERNE

40°

48°

80° 90° 100° 110° 120° 130° 140° 150° 160° 170°

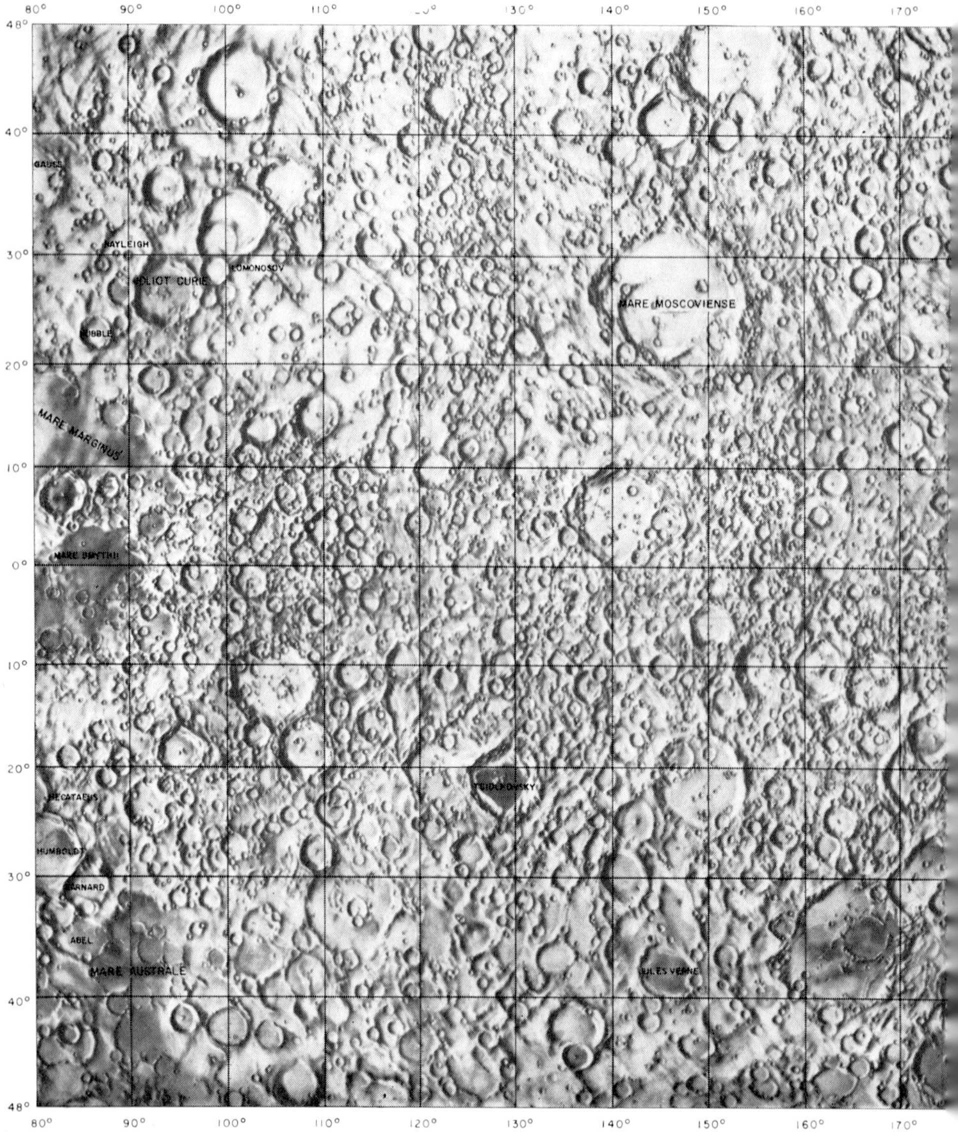

COMPILED FROM PHOTOGRAPHS, LUNAR ORBITER I, II, III, IV, V
(NASA 1966-1967)

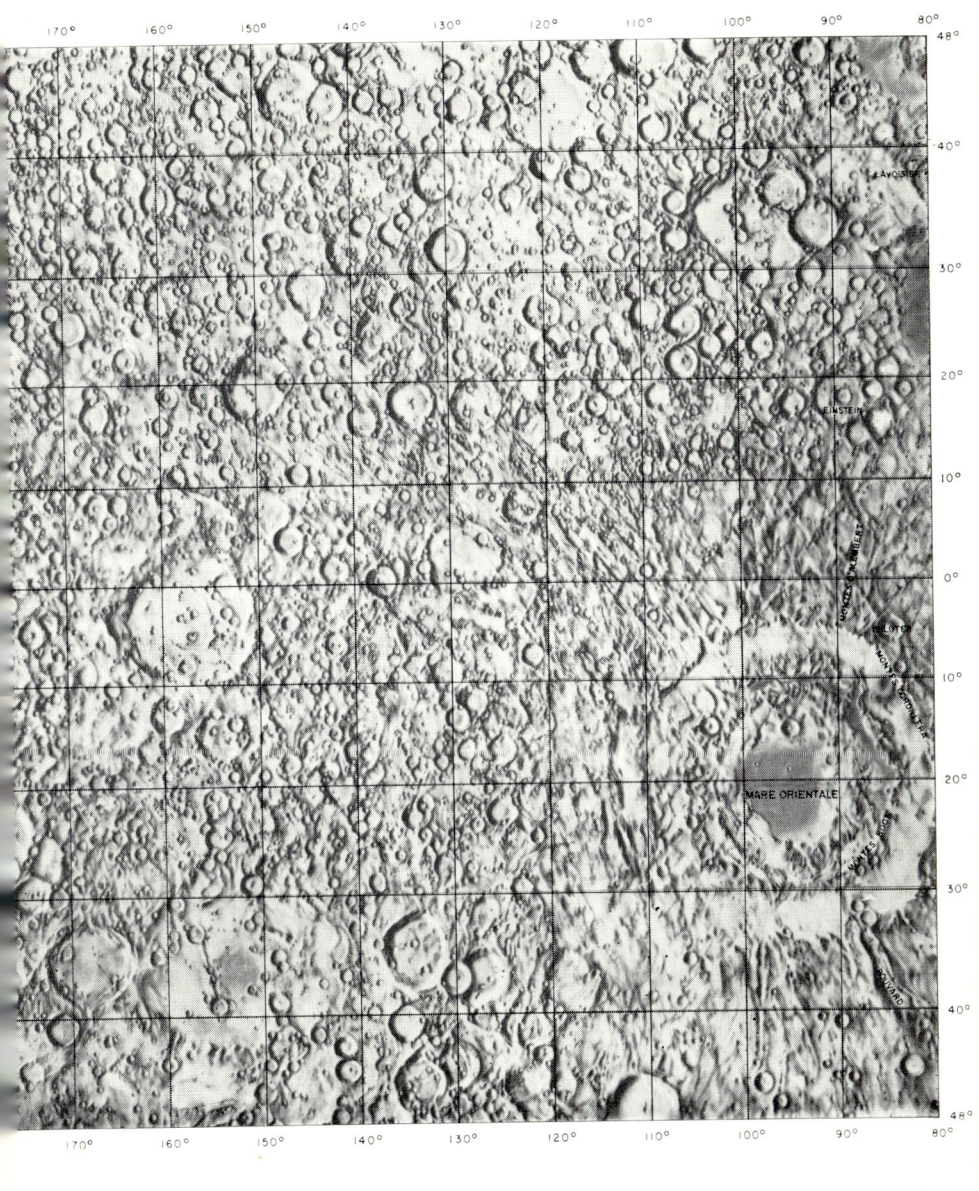

170° 160° 150° 140° 130° 120° 110° 100° 90° 80°

48°
40°
30°
20°
10°
0°
10°
20°
30°
40°
48°

LAVOISIER

EINSTEIN

MONTES D'ALEMBERT

DELISLE

MONTES ROOK

MARE ORIENTALE

MONTES ROOK

NOVARD

SCALE 1 20,000,000 AT 30° NORTH OR SOUTH

MERCATOR PROJECTION

ES

0 75 100   200   300   400   500   48°
24°
0°

100   200   300   400   500   600   700   800   48°
24°
0°

Plate 6c. This shows in fine detail the impressions in the lunar soil made by the *Apollo 11* astronauts. Part of the Lunar Module is silhouetted in the foreground. Michael Collins remained with the Command and Service Modules in lunar orbit while his colleagues explored the moon.

Plate 6d. A granular fine-grained iron magnesium rich rock. It appears similar to several igneous rock types found on earth. The sample was collected by Astronauts Armstrong and Aldrin during their moon walk on 21 July 1969.   *N.A.S.A.*

molecular dissipation. Both geology and selenology bear witness to this having occurred. The difference, however, between the Earth and Earth-like bodies and the Moon and Moon-like bodies, such as the large satellites of Jupiter and Saturn, is one of degree, the escape of the volatiles from the interior being much later and less complete in the case of the latter. Their structure to some depth below the surface must be rather like that of Swiss cheese, the holes being filled with gas and water (or ammonia), frozen or unfrozen, whence the low overall density. In fact, the cratered surface may be no more than an outward expression of this internal condition.

That the Moon's surface has been the scene of widespread and prolonged volcanic activity, during which, as we have seen (p. 53), a considerable shrinkage in volume may have occurred through the loss of imprisoned volatiles, is beyond any doubt.[2, 5, 8, 14] Whether or to what extent this activity has been associated with, or related to, meteoritic bombardment,[12] is of no great moment to our present quest. Similarly the issue of continuing vulcanism is beside the point, which is its effect on the present condition of the lunar crust, if this term can be applied to the Moon in its terrestrial connotation.

Here the difference in mass between the Moon and the Earth is once more of paramount importance, inasmuch as, firstly, it results on the former in much lower compacting gravitational stresses, and, secondly, leads to the virtual absence of any substantial atmosphere or hydrosphere through most of selenological history, although there exists some present evidence for a transitory, but nevertheless extended, period when liquid water appeared on the lunar surface.[6] On the whole, however, it will be correct to assume that the consolidating action of water and subaerial denudation have never been very effective.

Thus, even if the primitive surface has not passed through a molten stage, it will have been gradually overlaid, through long selenological ages, with layer upon layer of light, largely unconsolidated volcanics in the form of loose tephra, interlarded with and partly cemented by flows of lava or fluidized ash. It has grown from within, like the bark of a tree, at the expense of the hot interior.

I believe I was the first (1959)[4] to draw attention to the point that any lava reaching the sub-vacuum of the lunar surface would behave rather like the molten glass in Washburn's experiments, where it expanded under vacuum to 20 – 50 times its normal volume.[15] This high degree of vesiculation of lunar lava has now been generally admitted (*see* Appendix, p. 247) and confirmed by close-up views of the surface of the Moon. Experiments with terrestrial rock melts exposed to vacuum have been made by G. P. Kuiper and associates in the Lunar and Planetary Laboratory of the University of Arizona, and by G. Fielder's team at the University of London Observatory.[3] The former has obtained basaltic rock froths, or *reticulites*, having bulk densities of 0·1–0·2 and crushing strengths of about 3 kg/cm².[11] Lunar surface formations in the maria may consist of comparable material.[8]

A simple calculation will show that such reticulite will not be crushed by an overburden of itself less than 1 km thick. And being crushed does not mean that it will be compacted forthwith to the consistency of terrestrial basalt; its bulk density need be no more than doubled, which would still leave it below that of our pumice. Moreover, it the overburden contains large cavities, as may be expected, it will have to be appreciably over 1 km thick to overcome the resistance of surface-formed reticulite. Less expanded forms of reticulite, originating some way below the surface and so of a higher bulk density and crushing strength, must likewise be reckoned with, and these will be able to withstand the weight of several kilometres of lunar volcanics before giving way. Moreover, the escape of gas from underground intrusions must necessarily be less than from surficial effusions, so that the less frothy micro-structure will be partly compensated by the formation of larger drusy cavities.

We can, however, go beyond mere conjecture. A small piece of rock picked up by the scoop of *Surveyor III*, which landed in Oceanus Procellarum on 19th April 1967, would not crush under a pressure of 10⁷ dynes/cm². Putting once more their mean density at 0·2 and disregarding large cavities, such rocks will resist crushing down to at least $3\frac{1}{4}$ km and possibly more. In other words, bulk densities of lunar rocks should be substantially below unity for several miles below the surface, even without large-scale cavitation. Yet below this level rocks must still be expected to be drusy, and gas-filled hollows with walls of solid rock will be able to stand up to far higher pressures.

From this it will be seen that the differential pressures at the base of lunar mountains, and even the whole of the geometrical bulge, will make little call on isostatic adjustment and may be wholly absorbed by the structural strength of the rigid honeycomb crust itself. A 10 km mountain of mean density 0˙5 under a gravity of 1/6 g would correspond in terms of weight to a terrestrial hill about 1,000 metres high. The effectiveness of isostatic adjustment depends on differences of weight, so that it cannot be equally important in selenology as in geology. It also seems doubtful if mantle convection[2, 17] (see p. 44) could leave any strong mark on the surface features.

These points will become still clearer if we consider the likely internal structure of the lunar globe on the assumption that its nife core accounts for 10 p.c. of the total mass. As stated (p. 52), this leaves the rest of the Moon with a mean density of 2˙82.

Without any claim to finality, let us consider a three-layer model, in which the core is surrounded by a sima mantle of mean density 3˙30, sheathed in a peel of light porous rocks, with a honeycomb of caves, permeated with volatiles and having an average bulk density of 1˙00.

Simple calculations will show that on these assumptions the core will account for 35 p.c. of the Moon's radius, the mantle for 55 p.c., and the 'crust' for the remaining 10 p.c., or 174 km, the second layer comprising 71 p.c. and the third 19 p.c. of the total mass. The highly compressed nife core of the Earth occupies 54 p.c. of the radius, so that there is no reason to think that the figure taken for the relatively uncompressed lunar core is excessive; but any slight reduction in its size will react strongly on the residual depth of the 'crust', which appears enormous by terrestrial standards (Mohorovicic suggested a lunar crust 400 km deep). The separation of the metallic phase from the sima rocks should be less advanced than in the Earth, so that the Moon's mantle could be heavier than assumed in the bottom part, shading off into sial at the top. The exact consequences of this situation are not easy to estimate. It must, however, be borne in mind that in the conception proposed higher up the low-density layer of the Moon is no strict counterpart of the Earth's crust, which is neither a honeycomb structure, nor does it imprison the original allowance of volatiles to anything like the same extent.

The thermal evolution of the Moon [12, 17, 19] is important at this juncture.

Its first hot stage will be due to gravitational compression and evolution of energy from the short-lived radioactive isotopes, followed by a cooling-off period before the eventual slower reheating by the long-lived radioactive substances.

This second epoch would lead to further consolidation, affecting mainly the deep interior, and only imperfectly followed by the rigid honeycomb above. At a certain level between the two a gap may be expected to open and to be filled with water, forming a kind of underground sea. This period of internal contraction may be contemporaneous with the surface shrinkage inferred by Spurr from the distortion of the outlines of the old ring complexes. Partial collapse of the honeycomb at this or a later stage in what Spurr describes as 'the Imbrian revolution' would result in the formation of the maria and a renewal of volcanic activity.

The concentration of the maria on the Earthward side of the Moon clearly points to the association of this 'revolution' with the capture of the Moon by the Earth, probably between 1 and 2 æons ago.[7, 12] The collapsed areas of the crustal honeycomb would be densified, partly by the infusion of heavier material from underneath, partly by the pneumatolytic metamorphism of the shattered formations, and possibly by partial melting and degassing. Indeed, the *Surveyor* data do indicate that the marial lunabase is intrinsically heavier than the upland lunarite, while the orbital behaviour of the *Lunar Orbiter* craft has disclosed an excess of mass in this region (see p. 157). It seems, therefore, that the dynamic bulge (p. 42) arises not so much from any distortion of the Moon's shape as from the higher average density of the lunar crust where it faces the Earth. Since, though, the collapse of the honeycomb structure has been only partial even in this region, it will still have sufficient architectural strength to support its local gravitational anomalies without much buckling, while the mantle beneath it conforms to the figure of hydrostatic equilibrium.

REFERENCES

1 ALLEN, C. W. (1963). *Astrophysical Quantities*. 2nd Edn. University of London Press.
2 FIELDER, GILBERT (1965). *Lunar Geology*. Lutterworth Press, London.
3 FIELDER, G. *et al.* (1967). 'New data on simulated lunar material'. *Planetary and Space Science,* **16**, pp. 1,653–1,666.

4 FIRSOFF, V. A. (1959). *Strange World of the Moon*. Hutchinson, London.

5 FIRSOFF, V. A. (1961). *Surface of the Moon*. Hutchinson, London.

6 FIRSOFF, V. A. (1968). 'Water within and upon the Moon'. *New Scientist*, **37** (587), pp. 528–530.

7 GERSTENKORN, HORST (1967). 'The importance of tidal friction for the early history of the Moon'. *Proc. R. S., Series A,* **296** (1446).

8 GUEST, J. E. and FIELDER, G. (1968). 'Lunar ring structures and the nature of the maria'. *Planet. Space Sci.*, **16**, pp. 665–673.

9 *Handbook of the British Astronimical Association 1968.*

10 HÉDERVÁRI, PETER (1962). 'A theory about the internal composition and development of the Moon'. *Contributions from the Institute of Astrophysics and Kwasan Observatory.* Nos. 109, 126. Kyoto.

11 HESS, W. H. *et al.*, Eds. (1966). *The Nature of the Lunar Surface – Proceedings of the 1965 IAU–NASA Symposium.* The Johns Hopkins Press, Baltimore.

12 KOPAL, ZDENEK (1966). *An Introduction to the Study of the Moon.* Reidel Dordrecht.

13 KOPAL, ZDENEK (1967). 'The shape of the Moon, its internal structure and moments of inertia'. *Proc. R.S., Series A,* **296** (1446), pp. 254–265.

14 KUIPER, G. P. (1954). 'On the origin of the lunar surface features'. *Proceedings of the National Academy of Sciences of the U.S.A.*, **40** (12), pp. 1,069–1,112.

15 MACBAIN, J. W. (1932). *The Sorption of Gases and Vapours by Solids.* Routledge, London.

16 NESS, N. F. *et al.* (1967). 'Early results from the magnetic field experiments on Lunar Explorer 35'. *Jrnl. of Geophysical Research,* **72** (23).

17 RUNCORN, S. K., Ed. (1968). *Mantles of the Earth and Terrestrial Planets.* Interscience Publications.

18 SPURR, J. E. (1949). *The Shrunken Moon.* Business Press.

19 SUESS, H. E. (1965). 'Chemical evidence bearing on the origin of the solar system'. *Annual Review of Astronomy and Astrophysics,* **3**, pp. 217–232.

20 TURKEVICH, A. L. *et al.* (1967). 'VII. Chemical Analysis of the Moon at *Surveyor V* Landing Site: Preliminary Results.' *JLP Technical Report 32–1246.*

21 TURKEVICH, A. L. *et al.* (1968). 'VII. Chemical Analysis of the Moon at *Surveyor VI* Landing Site: Preliminary Results'. *JLP Technical Report 32–1262.*

22 TURKEVICH, A. L. *et al.* (1968). 'VIII. Chemical Analysis of the Moon at *Surveyor VII* Landing Site: Preliminary Results'. *Surveyor VII, A Preliminary Report, NASA SP–173*, pp. 207–233.

6

The 'Old Moon'

The traditional 59 p.c. of the lunar surface accessible to Earthling eyes deserves special notice, not only because it has been an object of sustained study ever since the invention of the telescope in 1609, but for its own sake, for it is a most remarkable part of the Moon. It is much more like the Earth than the rest. The familiar pattern of dark and bright areas, variously construed by popular fancy as a human face, a man with a dog, a rabbit, a donkey, or a crab, has no counterpart on the Moon's averted half. Except round the edges, it is substantially devoid of the dark areas, which Galileo thought to be seas and referred to as *maria,* accordingly, in contradistinction to the bright *terrae firmae.* And maria and terrae they have remained to this day.

Conventionally the old face of the Moon is divided into four *quadrants,* which are obtained by drawing at mean libration two straight lines at right angles to each other pointing north and south, and east and west respectively, through the centre of the lunar disk. The Moon was looked at chiefly as a celestial spectacle, so that its cardinal points were arranged to fit our sky, the west limb of the Moon looking towards our sunset and the east towards our sunrise horizon. Since, though, the rotation of the Moon is direct, from west to east, this meant that the Sun was falsely assumed to travel from west to east in the lunar sky. Moreover, an astronomical telescope usually inverts the image, so that the south pole of the Moon or a planet appears at the top, and a convention became established thus to present lunar drawings, photographs and maps upside down, with east and west reversed mirror-wise (Plate 1).

The astronauts of the Space Age found this intolerable, and began to draw maps of the Moon in the ordinary geographical way, with the north at the top and the east on the lunar sunrise horizon. This new look received the official accolade from the International Astronomical Union in 1961 and will henceforth be followed here. But old habits die hard, and some confusion persists. One of the legacies of the past is the name Mare Orientale, which means Eastern Sea, although now the Mare appears on the west limb of the Moon, unless perhaps viewed from behind — a new habit yet to be acquired!

However, to return to the quadrants, the First Quadrant is in the new North East, the Second in the North West, the Third in the South West, and the Fourth in the South East, going widdershins. In my *Moon Atlas*[8] (1961) I suggested dividing the far side of the Moon into a further four quadrants, repeating the procedure, the Fifth Quadrant being behind the Second, the Sixth behind the First and so on, but this suggestion does not appear to have caught on.

Owing to the librations, the true limb and the apparent centre of the Moon shift position, so that each quadrant is associated with a librational zone covering $2\frac{1}{4}$ p.c. of the averted hemisphere.

The polar regions and most of the Fourth Quadrant are closely comparable to the 'farside', the maria being concentrated within the first three quadrants. A string of minor maria follows the librational zone, especially in the east, and extends some way into the invisible part of the lunar surface.

Kurd von Bülow, Jack Green, and G. C. Amstutz[17] have proposed a revised 'geometric terminology' for lunar features, in which *telluric* is opposed to *lunar*, while *terrestrial* is regarded as an opposite of *oceanic*, but have overlooked the necessity of having some adjectives coresponding to the lunar *maria* and *terrae,* and the terms, *marial* and *terral* have been introduced here for this purpose. Generally speaking their terminology does not appear to be particularly felicitous and has not been adhered to in this description.

The terrae lie on the average about 2,000 metres above the maria, are relatively rough and pockmarked with innumerable walled and at times unwalled shallow hollows, which under small magnification make the Moon look rather like a thimble. The maria, on the other hand, are smooth by comparison and may be described as plains, while the terrae are also referred to as uplands or highlands, although some terral areas may actually lie below the level of some marial ground with reference to the centre of the lunar globe. The present knowledge of the figure of the Moon, however, is somewhat uncertain, so that such refinements are hardly justified, and in any case this picture is correct in broad outline.

Its resemblance to the Earth lies in that our deep-ocean beds are composed of dark basic rocks, while the continents and the continental shelves are made up overwhelmingly of light-coloured acid formations and the products of their decay. Thus, if the Earth were stripped of its waters and mantle of vegetation it would look to a distant observer not unlike the Earthward side of the Moon.

The basic rocks are heavier than the acid ones, which they generally underlie, so that this vertical distribution is due to isostasy (p. 16). We have seen that the *Surveyor* analyses indicate a similar relationship between the lunabase and the lunarite, the former containing more of the heavy metals. Yet three random samples are not enough for secure generalization; the differences of texture may be more important than those of chemical composition.

In any event we cannot be sure that all dark, or all bright, areas are of the same nature. For instance, there is a largish dark area extending like a stain over the marial terrain of Sinus Aestuum as well as some of the mountains dividing it from Mare Vaporum, without any apparent structural change in either. The dark colouring must, therefore, be due to some kind of superficial overlay. The bright haloes surrounding some craters and the crater rays, to be considered later on in more detail, are undoubtedly surface deposits, which are, moreover, very patchily distributed. It is, therefore, safer to regard the terms, lunabase and lunarite for the time being as descriptive of the surface colouring only.

On the whole the boundaries between the two types of surface are clear and unmistakable, often following geometrical lines, defined by fractures, ridges or sharp changes of gradient, which betrays a generic difference. Nevertheless, areas of intermediate shade exist. Not all lunabase is equally dark, nor lunarite equally bright. Lunar colourings are subdued, but there are suggestions of hue within the gradations of shade. On the basis of the latter alone samples of the lunar surface may be arranged in one continuous progression, with albedoes ranging from some 0˙03 to perhaps as high as 0˙40 in the core of the lightest features, although Aristarchus, which shines like snow in a full Moon, has an average albedo of barely 0˙20, which is less than granite and close to volcanic ash. N. N. Sytinskaya's investigations show hardly any overlap between the brightness diagrams of the lunar rocks, on the one hand, and terrestrial formations and meteorites, on the other.[6] The lunar rocks are generally darker. Since there is no marked difference in chemical composition between them and their terrestrial counterparts, this photometric disparity must be due mainly to differences of texture, the lunar surface formations being fragmented, porous or filamentary, as indicated in the foregoing chapter, and their reflecting surface including a high proportion of dark hollows. This is why a half Moon gives only about 8 p.c. of the light of a full Moon.

It has been mentioned that some of the darkest and brightest features represent surface deposits. This may apply to all selenologically recent fractures, which stand out as bright lines under a high Sun.[7] Crushed rocks are usually much brighter than the same rocks in compact state, but a lower porosity of the exposed fracture faces could produce a similar effect.

On the other hand, it is generally true, at least within one and the same formation, that the lowest portions are also the darkest and the highest the brightest, irrespective of whether the material is lunarite or lunabase.[7] Such differences show best at full phase, when the summits of the Apennines glow like a string of pearls against the lower slopes and most craters stand out as bright rings. The resemblance to the snowcapped mountains of the Earth may be fortuitous, but Martian calderas photographed by *Mariner IV* also show bright rims, which is attributed to hoarfrost, and hoarfrost need not be ice – it could be some readily volatizable salt. Its preferential deposition on the summits would, nonetheless, require the presence, at least at some time in the past or temporarily, of an atmosphere with a steep thermal gradient. Isostasy alone could hardly produce such an effect. Yet, whatever the explanation, the observation stands : the relative brightness of lunar surface features is related to altitude.

The early lunar cartographers followed faithfully the analogy with the Earth and gave a medley of fanciful names to both the terrae and the maria, but only those of the latter have survived. Thus we have a Mare Imbrium, or Sea of Rains, and Mare Nubium, or Sea of Clouds, both of which are wholly innocent of either rain or clouds, although, as we shall see in due course, these names may have some retrospective justification after all. However this may be, the absence of names for upland areas is awkward in description and necessitates reference to the mountains or ring complexes, which alone bear the names of terrestrial ranges and famous men respectively. The latter vary from the philosophers of antiquity to Kaiser Wilhelm I (no Hitler), but are mainly those of astronomers. The associated minor craters are designated by the name of the main ring followed by a Roman capital letter, e.g. Hesiodus A. The lower case is used for still smaller features. Some peaks are dignified with individual appellations, such as Pico, Mt. Hadley, or Mont Blanc, the highest mountain in the Lunar Alps. Failing this, the practice is to put a small Greek letter after the name of the range or group, as exemplified by Leibnitz β in the Leibnitz Mountains near the south pole. The same procedure is followed in the neighbouring Doerfels, some of which stand 10 km or more from foot to head, but the Himalayan system has been applied to the unclassified huge peaks which peep over the Moon's edge at favourable libration from beyond the pole, and these are called M1, M2, etc.[8] Small pointed bits of terra jutting out into a mare are sometimes referred to as *promontoria* (plural of *promontorium,* meaning promontory or cape), e.g. Pr. Heraclides, but 'islands' and 'peninsulas' remain unrecognized. Otherwise the rule is – when in doubt refer to a crater; to take Hyginus Cleft (or *Rille*) as an example, i.e. a groove-like depression associated with – in fact, incorporating – the crater Hyginus.

Both dark and bright areas or formations may be large or small, well defined or diffuse, and assume a wide variety of shape, but more often than not their outlines can be resolved into arcuate or linear elements, the circular or arcuate structure being typical of the smaller features, passing over into polygonal or linear as the scale increases. Even those large formations which look round at first sight become polygonal when the curvature of the lunar globe is taken into account. The most common polygon is a hexagon;

quadrilaterals and triangles are rare among walled enclosures, though frequent in ground structure.

As already mentioned, a hollow with upstanding rims is a recurrent theme of the lunar surface structure, but there are also *domes* and *tumuli*, cumuliform or botryoidal elevations (Plate 40), as well as upraised or depressed blocks of ground that may be tilted or dismembered, and various purely linear features of positive (ridges) or negative relief (depressions), or else superficial differences of colour, as in crater rays.

With few exceptions the material of the vertical relief is lunarite, even within lunabase country, while lunabase is typically associated with horizontal features, subdued sculpture, and at most low heights.

A rimmed depression may be described as a bowl, the Greek for which is *crater*, and this is how such formations are generally referred to, which implies no prejudice as to their mode of origin. No lower limit can be fixed to craters, as lunar rocks, being porous, abound in small cuplike depressions, which may be simply such pores exposed on the surface or have been caused by the impact of small meteoroids or any other agency.

Crater shapes vary a good deal. Some have ragged, irregular or scalloped edges; others are so neat and round as though they had been turned out on a lathe, which applies in particular to those from about 100-metre to kilometre range. The close-ups of the lunar surface in the region of Mare Nubium (crater Fra Mauro) obtained by *Ranger 7* in July 1964 first revealed the existence in this mare of multitudes of soft-shouldered, subdued depressions, as though worn down or overlaid by a later deposit. This feature reappears in the Alphonsus area, photographed by *Ranger 9* in March 1965, and the statistical investigation at the Goddard Space Flight Center in the U.S.A. has shown that these subdued depressions measure from some 10 to 300 metres across, while the smaller and larger craters stand out clear and sharp.[4] It has also been found that many of both the subdued and the sharply cut craters have funnel-like concave floors;[9, 11] sometimes, indeed, there is a distinct second hollow in the middle. This feature disappears more or less completely in rings over 10 km in diameter, though even smaller craters may have flat or convex floors. Small and medium craters with disturbed irregular floors also exist. Some of the smaller craters are practically devoid of a bounding wall and described as pits.

The complexity of structure increases with size. Typically a large ring complex, say between 20 and 100 km in diameter, has a multiple wall, which is terraced inwards and attains its greatest height at the outer rim. According to Fielder[6] the slope of the topmost terrace may be as steep as 50° inwards and at times as much as 30° outwards; the lower terraces are not so steep. These angles vary with the depth of the bowl and the elevation of its rampart above the surrounding country. On the outside, however, the slopes, or *glacis*, fall away gently at an angle decreasing from 10 to 5°. They are often humped, grooved or cratered, and where the nature of the surrounding ground permits (e.g. Aristillus) sub-radial shallow gullies may be traced under oblique illumination for tens of kilometres beyond the upswelling on which the crater sits, while bright rays, where present, may extend over hundreds and even thousands of kilometres in some preferred directions or all around. Sometimes they coalesce into a halo towards the crater, but bright or dark haloes need not be accompanied by rays. Radial dark bands are not infrequent in some well-developed ring structures, and at least one dark ray may be seen to the south-east of the 90-km (56-mile) crater Copernicus.[8]

On the inside the terraces of decreasing height may succeed one another with almost amphitheatrical regularity, as they do in Hell and Herschel (both of which medium-sized), but are more often interlaced. There are also concentric rings (e.g. Hesiodus A) and rings with spiral walls (Sabine). Often at the foot of the lowest terrace there is a kind of pediment, sloping down to the level central portion of the floor. The floor may be smooth, even in seemingly undamaged formations, such as Abulfeda, but characteristically one or more 'central peaks' or domes heave up from it. Such central peaks may be eccentric, and often have small craters at the summits. The peaks may be small, as in Pitatus, or fill up most of the interior, as in Alpetragius.

There exists a so-called Mädler's Rule, whereby the central peak is lower than the bounding rampart; it has few exceptions. It is also generally true that the floor of the crater lies below the level of the surrounding country, but Fielder lists four craters: McClure D, Ukert, Krieger and Taruntius, whose floors lie above this level, while the famous 80-km bowl of Wargentin in the Third

Quadrant is filled up nearly to the brim, so that the ring forms a kind of table mountain. There are smaller examples of the same structure.

The relative depth of the bowl and the height of the ramparts may vary a great deal in comparable craters, but statistically decrease with the size of the formation; the level portion of the interior behaves in the opposite way, and the largest rings are like vast plains, standing near the centre of which an observer would be unable to see the bounding walls. The walls themselves become like mountain ranges, often cratered along the rims. The largest of such *walled* or *bulwark plains* on the 'Old Moon' is Bailly, near the south-western limb in the Third Quadrant: it measures 295 km (185 miles) wide. Flat-floored craters 50 – 100 km in diameter are sometimes described as *crater plains*.

The floors of the great rings of Bailly and Clavius (230 km) are lunarite, but many of the walled plains, especially adjacent to the maria, have lunabase floors, best exemplified by Grimaldi (190 km), Ptolemaeus (145 km) and Plato (96 km). Such rings are often described as 'flooded' and their floors tend to be featureless.

A special type of formation which Fielder has called 'elementary ring' (Plate 4a) is substantially limited to marial terrain: it has no central depression and consists of a low undifferentiated annular mound, which may be complete or incomplete (Egede, Wallace, Stadius).[6, 10] Finally, we have *ghost rings*, which are marked out from the mare by a lighter shade alone.

If Bailly had a lunabase floor it may have been called a mare; indeed, it is much larger than the irregular lunabase area of Mare Undarum in the First Quadrant. But the 217-km (134-mile) Schickard is denied this honour, despite its dark interior; so, too, is Grimaldi, which is to all intents and purposes a small mare, for many maria are also ringed depressions, and Mare Humboldtianum consists of two overlapping dark walled plains[8] no bigger than Grimaldi.

Thus there is no great consistency in nomenclature. Minor lunabase areas may be called *lacus* (lake): there are two of them, close together in the First Quadrant – Lacus Somniorum and Lacus Mortis. If enclosed on three sides by lunarite, they are usually termed *sinus* (bay), or else *palus* (marsh). Possibly the *paludines* are partly separated from the governing mare by craters, mountains or other lunarite structures; but Palus Somnii is a lunarite area, distinguishable only by its golden-brown hue.

Morphologically the maria are of two main types: sub-circular or polygonal, and irregular. Mare Imbrium (2nd Quadrant), 1,300 km across and some 7 km deep near the middle in relation to its offshore parts is the premier 'sea' of the Moon and typifies the first division. It is, in fact, a very large walled plain, ringed by a mountain girdle locally exceeding 6,000 metres above the floor at its foot, and sloping gently away. The 300-km Mare Orientale is of the same type, ringed by a relatively higher and triple mountain scarp. Mare Nectaris may have been comparable once upon a time, but has undergone considerable development since its selenological initiation. On the other hand, the 460-km Mare Crisium, while clearly hexagonal and surrounded by high ground, lacks the characteristic outward-dipping mountain scarp; nor is this well developed around Mare Humorum. Such maria as Tranquillitatis and Foecunditatis are transitional between the walled-plain and irregular type.

The irregular maria also vary in appearance. The long strip of Mare Frigoris, northwards of Mare Imbrium, is a kind of lunabase valley *in excelsis*, comparable in all but scale to the similar strips of lunabase ensconced between the mountains that ring Mare Orientale and named in the days when the true nature of this remarkable feature was not appreciated Maria, Hiemis, Veris, Aestatis and Autumni. Maria, Anguis and Undarum adjacent to Mare Crisium are somewhat similar, though they are probably chains of lunabase-flooded craters. The librational maria of the east limb fit somewhere halfway between this class and Mare Crisium, while lunabase flooding reaches its fullest expression in Mare Australe, most of which extends beyond the utmost line of libration into the hidden hemisphere.

The huge Oceanus Procellarum, some 6 million square kilometres in area and so larger than European Russia, may be called the Pacific of the Moon. It spreads from the Second halfway into the Third Quadrant, and is clearly non-circular. It approximates to a quadrangular outline, most of its western shore closely hugging the limb. We encounter the same square cut in Mare Nubium and its northern part, now known as Mare Cognitum, but their western boundaries are not clearly marked owing to the confluence with the neighbouring lunabase plains.

Larger craters are scarce in the maria. Coastal rings seem to dip into the mare, where the rampart is much reduced in height or lacking, transforming the ring into a bay. In fact, fractures paralleling the 'coastlines' bear clear witness to the maria being areas of repeated subsidence. A feature peculiar to the maria are *wrinkle ridges*, the highest of which and the only named one, the Serpentine Ridge of Mare Serenitatis, attains 1,000 metres, but the typical height is not more than 200 metres or less. Wrinkle ridges are often associated with surface fractures, but located inwards of them within the maria. They wind irregularly, like twisted ropes or crumpled skin on boiling milk. But the Serpentine Ridge is largely composed of arcuate segments, as though fragments of undeveloped elementary rings, which is also true of some wrinkle ridges in the neighbouring Mare Tranquillitatis. At times a wrinkle ridge is cracked open at the top (Kuiper), the crack being brighter than the rest. They also go together with domes, which are low swellings of the ground, often surmounted by a small crater, generally of a circular or sub-circular outline. Domes are particularly well developed in the area of the crater Hortensius and elsewhere in the Oceanus Procellarum, but they also occur inside craters (e.g. Aristillus). The average diameter of a dome is 15 km. Many are so low as to cast no perceptible shadow even under a low Sun. The *Orbiter* photography has shown them to be among the commonest features of the lunar surface.

Tumuli are similar but much steeper hills, while *Orbiter II* has spotted some steep projecting rocks in Mare Serenitatis. Tumuli are frequently of a triangular or crescentic outline, or else elongated in conformity with the lie of the land. Such lone mountains as the 1,300-metre Pico and its neighbour Piton in northern Mare Imbrium may be described as very large tumuli. Tumuli run riot in some areas, for instance in the hinterland of the Lunar Alps, which themselves consist largely of scattered, often cratered, elongated heights, coalescing into a distinct chain, rather than a tilted scarp, in the south-east, and attain about 4,000 metres in Lunar Mt. Blanc. Their *Orbiter* views compose some of the most terrestrial landscapes to be found on the Moon (Plate 16).

A striking feature of the Alps is the Great Alpine Valley, a lunabase trough some 120 km long and up to 16 km wide, which splits the mountains in half as if they had been slashed by a gigantic sword.

The Alps are part of the much-breached mountain girdle round Mare Imbrium. After pursuing an ENE trend, the girdle breaks off abruptly in Cape Agassiz, to be resumed, with a 200-km eastern offset, by the meridional range of the Caucasus. This consists chiefly of shattered blocks, tilted away from Mare Imbrium and dropping towards it in a scarp of varying steepness. But, superimposed on this topography, a coastal range of crater-topped peaks attains locally some 7,000 metres. After a gap, about 100 km wide, where Mare Imbrium coalesces with Mare Serenitatis, the girdle reappears in the magnificent range of the Lunar Apennines, which runs almost straight for 1,000 km from Cape Fresnel to the crater Eratosthenes. Mt. Huyghens, the highest point of the range, stands 6,600 metres above the level of the mare (Plate 2).

The character of these mountains is similar to that of the Caucasus, or of the 7,000-metre Rook Mountains to the east of Mare Orientale. In fact, these *chain mountains* may be compared to the ramparts of large walled plains, such as Clavius; but there is a difference, as the latter do not as a rule drop inwards in a single scarp and run down instead in several 'terraces' of declining altitude. The Apennines have residual foothills on the Mare side that may be construed as a similar architecture drowned in the mare. The Rooks have a well-developed lower terrace, but are separated from it by a comparatively wide and deep trench. However, single-scarp ramparts do appear in some 'flooded' craters with dark floors, such as Plato.

Although these mountain ranges bear a seeming resemblance to our Alpine mountains, this is misleading, as they have not been produced by folding and moulded by water and ice erosion. They are basically tilted blocks, shattered by tectonic movements. This character emerges very clearly in the Carpathians on the south shore of Mare Imbrium, while in the north-east the Jura Mountains above Sinus Iridum have escaped dismemberment and loft up in a fairly compact 7,000-metre scarp. The Altai range in the Fourth Quadrant is another example of *scarp mountains,* and the Cordilleras of the Orientale system may be put in the same category.

The high ground round Mare Crisium remains substantially untilted and is broken into segments by a grid of valleys like the 'monadnocks' of the Scottish Highlands. The Taurus Mountains to the north-west, bordering on Mare Serenitatis and Lacus

Somniorum, form another example of such *block mountains*. I
have previously described chaotic jumbles of tumuli, characteristic
of the Alpine and Caucasian hinterland, as *jumble mountains*. [7, 8]

These various types shade into one another, and it is not always
easy to decide which is which. The Leibniztes and the Doerfels,
however, referred to on p. 41, belong to a distinct type of *cellular
mountains*. They are still insufficiently explored, despite the
Orbiter surveys, but their prevalent pattern is clearly rectangular
and defined by successions of very large, squarish ring complexes,
whose walls run together into chains, the high peaks rising at their
meeting points. Transverse chains of smaller craters and rift
valleys contribute a further element to the scenery. [8, 16]

Both straight and arcuate chains, short or long, of small and
medium-sized craters are not uncommon in the Third and Fourth
Quadrants, even if we exclude such concatenations of walled
plains as Ptolemaeus-Alphonsus-Arzachel-Regiomontanus-Pur-
bach, or Langrenus-Vendelinus-Petavius-Furnerius. There are
also distinct crater pairs, mainly in meridional array, exemplified
by the Aristoteles-Eudoxus, Aristillus-Autolycus and Agrippa-
Godin pairs. The two paired craters are closely comparable in
structure and the (first) northern one is always the larger of the
two. More complicated crater sequences exist (Fig. 3), [6, 7, 13] but
one of the most characteristic features of the Moon are extended
chains of closely-packed craterlets, which may coalesce into one
continuous 'cleft', pass into a simple surface fracture, or a wrinkle
or other type of ridge.

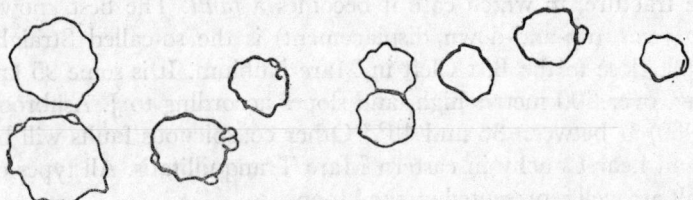

Fig. 3.　Repetitive pattern of craters in the Tacitus-Aliacensis group.

There is a conspicuous crater chain of this nature between
Eratosthenes and Copernicus. The craterlets are of kilometre
size, and cast long shadows under a low Sun, so that they are not
just 'blowholes' as Spurr thought. The chain is somewhat irregular
and joins in the south and north into valleys with scalloped edges,

a tell-tale feature of many similar depressions. Northwards the line of the chain is continued by a wrinkle ridge in the middle of Mare Imbrium.

The Birt Cleft, associated with a crater of this name, in the south-east of Mare Nubium is comparable, but more compact. The crater chain-cum-cleft in the Alps, north-east of Plato is again not unsimilar, but sinuous, which may denote a different origin. But there are also linear crater chains, which resemble the mark made on a sheet of paper by an unthreaded needle of a sewing machine. A good example of this structure will be found on the north-east wall of Ptolemaeus; there is another near the crater Opelt in Mare Nubium, and quite a few in the highland regions of the Fourth Quadrant, especially west of the Altai Mountains, where some chains are double with squarish, filled craterlets packed very close together. Short crater chains are frequently found on the outer glacis of such rings as Copernicus.

Although the object of this chapter is to avoid interpretation and concentrate on purely morphological description, the latter will be greatly simplified by the introduction of a few geological terms. Thus a narrow surface rift, sealed or unsealed, may be described as a simple *fracture*. There is a whole family of these radiating from the elementary ring of Letronne in the Third Quadrant, and another is associated with Gutenberg on the west shore of Mare Foecunditatis.

More often than not, however, either vertical or horizontal movement, or both, of the land has occurred on the two sides of the fracture, in which case it becomes a *fault*. The best known *dip fault* (up-and-down displacement) is the so-called Straight Wall, close to the Birt Cleft in Mare Nubium. It is some 95 km long, over 300 metres high, and slopes according to J. Ashbrook (1960) at between 36 and 48°.[2] Other conspicuous faults will be found near Cauchy in eastern Mare Tranquillitatis. All types of fault are well represented on the Moon.

Another frequent linear feature is a double fracture, where a strip of ground has subsided between the parallel banks. This is known in geology as a *graben*, and the opposite configuration where a strip of land is uplifted between the fractures is called a *horst*. Horsts likewise occur on the Moon, but not all, nor even the majority of, linear lunar ridges can be described as horsts. On the other hand, a typical lunar *cleft*, also known as a *rille* or *rill*, is definitely a graben. In fact, neither cleft, nor rill is a very suitable descriptive term, the one suggesting an abyss and the other a flow of water, which is not at all what Johann Schroeter, who was the first to describe these features late in the eighteenth century, had in mind when he used the German word *Rille*, meaning 'groove'. Patrick Moore suggests 'trough'.[13] Not to complicate matters, I will continue to use *rille* in its German spelling – after all, graben is also German for ditch, and ditch is perhaps the best alternative.

Graben-rilles are extremely common in many regions of the Moon, especially where lunabase and lunarite meet. A whole bunch of them crosses the walled plain of Riccioli and is beautifully displayed by the *Orbiter IV* frame 173, 173H (Plate 18). These rilles are shallow and relatively broad (up to several km) depressions, passing impartially over high and low ground without change of direction or appearance, and so evidently deep-seated. They look very much like the marks a heavy wheel would make in rolling over a hardened cake encasing a plastic substratum. Only occasionally is a rille blocked or bridged by a later formation. At times the marginal fractures are clearly visible, and may be outlined by small craterlets, as in the Moltke Rille, and the central strip may be slightly upraised or bulge up, as was first noticed by Fielder[5] (1956), when he studied the Ariadaeus Rille at Pic-du-Midi. The rims of some graben-rilles are somewhat elevated, as will be clearly seen at the foot of the Apennines (*Orbiter IV*, Frame 109, Site 20C).

Sometimes associated with graben-rilles, but also occurring independently, are broad shallow depressions of uniform depth and irregular outline. There are several such in Riccioli (interlinked) and at the foot of the Apennines. Like features in reversed relief are rarer.

Yet not all rilles are of the graben type : some are relatively deep, with triangularly concave floors, as the branched rille east of the Caucasus in Mare Tranquillitatis – also in Riccioli and elsewhere; or else a graben-rille has a ditch at its bottom. One type may pass into another, or into a simple fracture, which shows a common mechanism of formation.

All these rilles are unaffected by surface relief. Not so, however, the *sinuous rilles,* which usually originate in an elongated depression or crater, and meander intricately downslope, typically avoiding all elevations, growing thinner and shallower as they progress, and eventually peter out. The Hadley Rille at the foot of Mt. Hadley (Apennines) is a well-known example of this formation. There are more of them in the Alps and in the region of Aristarchus, where the great Schroeter Valley is somewhat atypical and has a secondary meander on its floor. It is comparable in some ways to the Hyginus Cleft, which is partly a graben and partly a crater-rille like the Birt Cleft.

We may recall that the structure of craters alters as they increase and the same is true of graben rilles. To the north-east of the great walled plain of Janssen, which measures some 180 km across, there is a great graben, 25 km wide and 185 km long, known as the Rheita Valley after the large crater at its northern end. It is a formation basically similar to the Hyginus Rille, much encroached by craters, some of which are prior to the subsidence and have sunk with it, while others, including Rheita itself, are posterior to it and undisturbed. The Valley ends up in a confluent crater-valley with a slight change of direction. There are further similar valleys near the south pole,[8] but they are difficult to observe owing to extreme foreshortening.

The country south-east of the Apennines, near Ptolemaeus and beyond has a peculiar appearance, rather like the bark of a chestnut tree, owing to the multitude of grabens, horsts and linear ridges, all fanning out away from Mare Imbrium, with some variation in direction. The system includes largish, but short crater-valleys, as well as peculiar rectilinear rilles, most of which have scalloped edges, sometimes passing into crater chains. Fielder has called them 'groove valleys'[6] and I, 'tear valleys'.[7] Unlike the graben-rilles, they avoid crater floors and clearly depend on surface relief, are interrupted by craters and change direction in climbing over crater walls; in other words, they are not as deep-seated as ordinary graben-rilles and have not involved any noticeable subsurface deformation. On the other hand, many faults, not excepting the Straight Wall, are aligned with them. To some extent their prominence is due to orientation: being nearly meridional, they cast clear shadows under both the morning and evening Sun, whereas the intersecting fracture swarms are illuminated lengthwise and do not catch the eye.

Some reference has already been made (p. 68) to the crater rays and haloes, as well as the *dark radial bands* which appear under a high Sun in such craters as Bullialdus and Aristarchus, and in the latter case extend considerably beyond the crater walls.[1] Some of these bands may be due to chains of dark-haloed craterlets, and at least some of the bright rays show a similar association with bright-haloed craterlets and surface fractures (e.g. McCall, 1966).[12] The exact relationship between the two types of formation, however, has not been firmly established, and in this one case the close-up photographs have been less revealing than one might expect for two reasons: the rays stand out best under vertical illumination when the details of surface relief are largely lost owing to the lack of shadows, which makes this situation unfavourable for general-purpose photography; and it is well known that faint extensive differences of colouring, especially when made up of fragmentary elements, are best seen from the distance. The rays are precisely that.

At full phase the extensive ray system around the 90-km Tycho, with individual rays over 1,000 km long, makes the Moon look somewhat like a peeled orange (Plate 1). This is why Tycho has been described as 'the metropolitan crater of the Moon'. The next most extensive ray system is that of Copernicus, though single rays emanating from Furnerius may be longer. Under oblique illumination rays fade rapidly as the angle decreases, and are said to disappear altogether when it drops below 8° (W. H. Pickering, A.P. Lenham). There is, however, some asymmetry in this respect, as I have observed telescopically and seen in photographs rays right on the terminator in a waning crescent, but never in a waxing one.[7] This fits in with the finding of Jan van Diggelen that bright haloes do not reach their maximal brightness at zero phase, but some 10° after.[15] It may also be noted here that ray systems look just as bright at the centre of the limb at quarter phase as they do near the centre of the disc at full.

There is a marked difference between the system of Tycho, which typifies terral craters, and that of Copernicus, characteristic of the maria. The tychonic rays run more or less straight, are more coherent and generally longer, as well as brighter, passing impartially over high and low ground. The rays of the copernican habit are ragged, tangled, crossing over one another, tend to form loops and comet-like *plumes* with the brighter narrow end facing towards the crater. They are impeded by quite low obstacles, such as wrinkle ridges. Spurr has compared their general pattern to 'that of snow thinly drifted by a strong wind across a black frozen lake'.[14]

The marial rays are also darker than the terral ones, which he sought to explain by the former being made of basic and the latter of acidic volcanic ash. Yet it will be observed that even bright tychonic rays lose their brilliance upon reaching a mare (Firsoff, 1961),[7] so that this behaviour may be due to the nature of lunabase and not of the rays themselves. There are also differences within the copernican type, the rays of Aristarchus and Aristillus being darker than those of Copernicus and Kepler, while the rays emanating from Theophilus, though traceable right across Mare Tranquillitatis, are exceedingly faint and grey.

Some of the rays trend more or less into the 'parent crater', others miss it by a good margin in the copernican habit, while tychonic rays are often tangential to it, which is well illustrated not only by Tycho itself, but by Proclus, Thales and Anaxagoras. In many cases, though in alignment with the main system, the rays clearly originate in small bright craters encountered on the way.

This completes our somewhat cursory general survey of the characteristic formations of the traditional lunar surface, but some of them will be considered in more detail at a later stage.

REFERENCES

1 ABINERI, K. W., and LENHAM, A. P. (1955). *J.B.A.A.*, **65** (4).

2 ASHBROOK, JOSEPH (1960). *P.A.S.P.*, **77**, pp. 55–58.

3 BLAGG, MARY A., and MULLER, K. (1935). *Named Lunar Formations*. Lund, Humphries & Co., London.

4 CAMERON, WINIFRED S. (17th September 1965). Private communication.

5 FIELDER, GILBERT (1956). 'Note on the bottom of the Ariadaeus Cleft', *Astronomical Contribs. from the University of Manchester*, Series III, N. 30.

6 FIELDER, GILBERT (1965). *Lunar Geology*. Lutterworth Press, London.

7 FIRSOFF, V. A. (1961). *Surface of the Moon*. Hutchinson, London.

8 FIRSOFF, V. A. (1961). *Moon Atlas*. Hutchinson, London, and (1962), Viking Press, New York.

9 FIRSOFF, V. A. (1967). *J.B.A.A.*, **77** (2) pp. 106–111.

10 GUEST, J. E., and FIELDER, G. (1968). *Planetary and Space Science*, **16**, pp. 665–673.

11 KUIPER, G. P. (1966). 'The Surface Structure of the Moon' in *The Nature of the Lunar Surface, Proceedings of the 1965 IAU-NASA Symposium*. Johns Hopkins Press, Baltimore.

12 McCALL, G. J. H. (1966). 'The concept of volcano-tectonic undulation in selenology' in *Advances in Space Sciences,* **8**.

13 MOORE, P., and CATTERMOLE, P. (1967). *The Craters of the Moon*. Lutterworth Press, London.

14 SPURR, J. E. (1944). *Geology Applied to Selenology – The Imbrian Region of the Moon*. Science Press.

15 STRUVE, OTTO (1960). *Sky and Telescope,* **20** (2), pp. 70–73.

16 WHITAKER, E. A. (1954). *J.B.A.A.,* **64** (6), pp. 234–242.

17 VON BÜLOW, K., GREEN, J., and AMSTUTZ, G. C. (1963). 'A Geometric Terminology of the Structural Features of the Moon', *Eclogae geologicae Helvetiae,* **56** (2), pp. 853–859.

7

The 'New Moon'

The invisible, 'forbidden' part of the Moon was a standing challenge to human imagination. Hansen's bulge (p. 38) created a brief excitement, kindling a hope of air, water and life out of our sight, at the blunt end of the 'lunar egg' – though, chiefly in the field of science fiction. There were, however, some sound reasons to suspect that the Moon's 'farside' may be different. For one thing it could never be eclipsed by the Earth. Lunar eclipses involve violent fluctuations of the surface temperature. Thus **Edison Pettit** observed on 28th October 1939 the temperature of a small area near the centre of the disk to drop from 159° C to −75°C within a single hour upon entering the umbra, and climb back to the original +97°C even faster when the eclipse had ended.[9] Such thermal changes must be accompanied by an intense volume expansion and contraction of the rocks, which – it was argued – would give rise to heat erosion, reducing the rocks to splinters, grit and eventually fine sand. Temperature varies much more gradually at sunset and sunrise, so that the absence of eclipses would greatly reduce heat erosion on the averted hemisphere, its relief being, consequently, better preserved and more rugged. There would also be a difference in the intensity of the body tides caused by the Earth, especially if the Moon once moved closer to it than at present.[3]

To what extent these expectations have been confirmed we shall see later on.

At the end of the nineteenth century (how remote it seems!) the

American geologist N. S. Shaler had the idea of locating invisible craters from the convergence of their rays coming from behind the limb. This work was subsequently taken up in Britain by Emley, Wilkins and Moore who between them produced a provisional chart of a belt of ground lying beyond the libration zone. The chart included a bright rayed crater, comparable to Tycho and approximately antipodal to it – in fact, a kind of 'anti-Tycho'.[11, 12]

This is what things looked like in 1954, but five years later, on 7th October, mankind was listening on their wireless sets with bated breath to the weak 'bleeps' emitted by *Luna* (or *Lunik*) *III*, launched 3 days earlier, as it swung behind the Moon and sent the first television pictures of its other side back to Earth. These pictures were exceedingly bad, but, coming less than a month after *Luna II* first hit the Moon, their very imperfection added to the sense of drama by preserving part of the mystery.

The Moon was then a waxing crescent, some 60° fat, and as seen from behind appeared in a complementary gibbous phase. This meant that most of the unknown ground far back was under near-vertical illumination and so unfavourably placed for viewing. The *Luna* pictures showed a wide strip of the known half as far as Mare Tranquillitatis, the eastern librational maria, and, beyond these, a largely blank expanse, with a few nearer craters faintly discernible and some dark patches of lunabase more clearly defined among indistinct shadings.

After some prolonged processing, involving techniques of selective image intensification, the Soviet Academy of Sciences brought out in 1960 the first map of the Moon's 'farside', with a sprinkling of new names and some doubtful identifications.[1, 5]

A long pause ensued, marked by a number of Russian and American failures. In 1964 and 1965 the American *Ranger* craft crashed into the Earthward hemisphere, yielding in their short operational lifetime the first close-ups of the lunar ground of superb resolution, far beyond the wildest dreams of the telescopist. Not until 20th July 1965, however, did another Russian probe *Zond III* emulate the feat of *Luna III* in obtaining further views of the Moon's back.[6]

The resolution was improved tenfold, the smallest features shown with reasonable clearness measuring about 3 km across. One of the best and most instructive pictures sent back by this

probe was of the opposite side of the averted hemisphere, around
and west of Mare Orientale. It gave the first intimation of this
surprising formation, previously seen only in great foreshortening
near the limb at favourable libration. The picture also revealed
numerous chains, some of them hundreds of kilometres long,
radiating from the Mare and composed of medium-sized and even
quite large (up to 60 km wide) craters. Curiously enough, it was
precisely the slight imperfection of vision that, by blurring the
outlines of the less clear-cut rings, threw these chains into sharp
relief. Another important finding of *Zond III* was the existence
on the 'farside' of huge walled plains, attaining and exceeding 400
km in diameter, and so comparable to maria, but with bright
lunarite floors. The Soviet scientists proposed for these rings the
name of *thalassoids* (from the Greek θαλασσα, sea).

This Soviet triumph in the space race, however, was to be
short-lived, as the Americans had in the meantime mastered the
technique of putting a probe into a circumlunar orbit by remote
control and of varying its altitude above the surface of the Moon,
and despatched between August 1966 and August 1967 five successive *Lunar Orbiters*, every one of which was a resounding success.

The photography was magnificent, covered both sides of the
Moon in great detail with the precision of an aerial survey of the
Earth. From these photographs was compiled a *Lunar Farside
Chart*,[7] published by the NASA late in 1967 and reproduced here
(Plate 6). The *Chart* is described as provisional, and may require
minor adjustment here and there, but its only doubtful feature is
the washy area, about 40° of latitude long and up to 20° of longi-
tude wide, in the extreme west of the south polar region.

It is interesting, if perhaps a little unfair, to compare this chart
with the original Russian map.

The resemblance is none too striking. Tsiolkovsky is unmistak-
ably there, and so is Mare Moscoviense. Of the Russian-named
craters Joliot-Curie reappears, coincident in position, but not in
shape, while its neighbour Lomonosov, while still listed in the
American map, is so ill-defined as to be hardly worth naming.
Most disappointingly, the bright-rayed 'anti-Tycho', christened
Giordano Bruno by the Russians at the latitude of approximately
37° N and longitude 103° E, corresponds to nothing readily
identifiable in the *Orbiter* pictures and the resulting chart. There

is still hope here, as the official *Lunar Orbiter Photo Indices*[8] seem to miss Bruno by a narrow margin. If, however, it were a real 'anti-Tycho' at least its rays ought to show in some pictures, but, having examined all of them, I could not trace any rays attributable to it.* This may be due to oblique illumination under which the photographs were taken, but the rayed craters of the Earthward side stand out quite clearly, and there just does not seem to be anything comparable on the other side. There are some craterlets with bright haloes and many crater walls are tipped with white,[10] but I could find only two or three small craters with short ray systems on the far side of the Maria, Marginis and Australe. Mare Orientale is a selenologically young formation. There are several well-developed craters with central peaks and resembling Tycho in appearance on the western edge of the Mare and beyond it on the 'farside'; these craters are certainly posterior to the Mare, but they are rayless. This is confirmed by the *Zond III* view of the area, although it shows several ray centres east of Mare Orientale.

The Soviet Range is a definite casualty. South-east of Tsiolkovsky, the dark-floored Jules Verne has survived, but all that remains of the putative Sea of Desire, *Mare Desiderii,* long suspected to exist beyond the Leibnitz and Doerfel Mountains on the principle of compensation,[3, 11] is a 275-km walled plain, which is partly lunabase, followed by several smaller flooded crater-plains. These, however, fail to coalesce into anything like a mare, if only of the underdeveloped type of Mare Australe.

Since then the Russians have compiled independently, under the direction of Yu. N. Lipsky a new map of the averted hemisphere, based on the results of *Luna III* and *Zond III*. It is incorporated in a *1:5,000,000 Map of the Entire Moon,* and is available also in the form of a school globe, the latter of which has an English version. This Russian map is less complete in coverage than the *NASA Farside Chart,* but, blank spaces apart, most of the main formations are readily identifiable in both. Curiously enough, however, their co-ordinate positions differ systematically in the two maps by several degrees in latitude and longitude. Since the American *Orbiter* photographs are superior to those taken by the Soviet probes, the *Farside Chart* is probably

* One of the colour photographs taken from *Apollo 8* in December 1968 shows clearly a crater with a vast ray system corresponding in position to Giordano Bruno.

the more reliable of the two, except possibly in one respect: the American cartographers have been rather shy of showing crater chains, prominently displayed in the Russian map, and tended to depict them as ill-defined valleys. I suspect that this may reflect a bias in favour of the meteoritic hypothesis of crater origin, which does not provide for the existence of crater chains.

In exercising their right of priority, the Russians have named 150 'farside' formations after scientists and thinkers of all nations, with a generous sprinkling of Russians, many of whom are quite unknown to me. This christening spree, however, must be considered provisional, as it has been agreed to postpone the final naming of the newly discovered features till the 1970 congress of the International Astronomical Union. In the latest editions of the *American Farside Chart* 486 craters are identifiable by Arabic and the thalassoids by Roman numbers. (See *Sky and Telescope*, **36** (3), pp. 147–151. September 1968).

This scarcity of lunabase and the total absence of proper maria is the most striking characteristic of the lunar 'farside'. There are, of course, the librational maria along the west and east limb, but they are borderland formations, partly visible from the Earth. Mare Moscoviense is the only partial exception to the rule. Located between 18 and 33° of northern latitude and 137 and 154° of eastern longitude, it is a small crater-mare, some 300 km in diameter, and so comparable in size to the central plain of Mare Orientale, but devoid of the latter's peripheral structure. Indeed, Mare Moscoviense resembles in every respect a large walled plain, only a part of its floor is lunabase, and its dimensions are exceeded by the largest thalassoids.

The latter are concentrated towards the poles. Some are clear-cut, with well-developed ramparts; others poorly defined, as though half-baked. There are also a few vaguely outlined ring structures, which appear to have been overlaid and effaced by later craters. Tsiolkovsky, which is unquestionably the premier crater of the 'farside', sits at the south end of one such residual or emergent thalassoid, some 500 km in diameter, in which it compares to Schiller on our half of the Moon. A semi-marial, unwalled lunarite plain of roughly the same size extends south-eastwards of Joliot-Curie.

Tsiolkovsky is a truly magnificent formation, without a peer on either side of the Moon. Somewhat irregular in outline and running into a point north and south, it measures 190 km wide from wall crest to wall crest. The ramparts tower at least 7,000 metres over the heart-shaped 'lake' of dark lunabase, thinly peppered with craterlets and shored by wrinkled lava. They are finely scarped and terraced, and very steep, especially in the west (Plate 14). A mountainous volcanic island rears up from the 'lake' a little eccentrically and sports a triangular peak with a blocked summit vent. The whole vividly recalls Crater Lake, in Oregon, but the latter's cradling caldera is only 13 km wide, and so puny by comparison.

The combination of a smooth lunabase floor with a central mountain is most unusual, and it is equally unusual to see so well-developed a central mountain in so large a ring.

In the south-east the bounding wall sends forth five fingerlike 'piers', which compose a kind of paw, while on the north-east glacis unmistakable flows of lava descend into the lowly neighbouring craters. Evidence of volcanic action is plentiful all around. Southwards a short valley connects Tsiolkovsky with a nameless 80-km ring, whose floor is a spit image of an Icelandic lava field, flood upon flood having surged gently upwards in thin, superimposed sheets.

Yet the unique feature of Tsiolkovsky are the two curious aprons on its west side. These are marked by continuous longitudinal striations, nothing like the surfaces of a lava flow, as well as transverse fractures, and must be at least 100 metres thick at the head, ending up in a steep wall. The north-western apron, which is the larger of the two is over 90 km wide and some 85 km long at maximum extent, neglecting the effect of the slope, which will not exceed 5°, except at the very top. The apron is clearly bounded by a ditch above and on the north flank, and mimics to perfection the aerial appearance of some glaciers.

Oddly enough, the *Lunar Farside Chart* does not show even the larger apron. Nothing so shocking is allowed on the Moon!

The same Frame 121 of *Orbiter III* which centrally features Tsiolkovsky shows three thalassoids spaced along the eastern edge of Mare Australe. Southwards the Mare degenerates into single flooded craters and is bordered, somewhat chaotically, by high unmeasured mountains. The mountains stand in clumps and short ridges, in partial coincidence with crater walls, and in a vaguely orthogonal pattern, as do the Leibnitzes farther west in another frame. In fact, they both belong to the cellular class.

The summits of the 'Australian' mountains are invariably cratered, and their steep conical slopes suggest volcanism of acid type, exemplified on the Earth by Vesuvius. In a few cases the small summit crater is replaced by a wide and deep-trenched caldera, 30 – 50 km in diameter, whose floor is raised high above the surrounding country. This again differentiates them from the usual run of lunar craters and assimilates them to terrestrial volcanoes.

Orthogonal lineaments, which we shall have occasion to discuss in more detail, are not peculiar to the polar regions, though often strongly marked there, or to the averted hemisphere. In the Third Quadrant, not far from the centre of the disk, there is a rectangular walled enclosure between the craters, Palisa and Davy. It measures 38 by 55 km. But the 'farside' boasts a true rectangular walled plain, some 300 km long and 140 km wide, just over the visible limb, beyond the craters, Alston and Ulugh Beigh, and adjacent to the Hercynian Mountains of the old maps. These mountains are a somewhat doubtful item, not readily identifiable on the *Farside Chart*,[7] which includes a 10° overlap with the Earthside, any more than are the d'Alembert Mountains north of Mare Orientale, the former being most probably the high ramparts of the great rings associated with the rectangular walled plain.

Among the linear, or sublinear, features mention has already been made of the long crater chains extending north-westwards of Mare Orientale, whose main part, though visible at favourable libration, lies beyond the 90th meridian. This mare with its triple ring of high mountains, at least 600 km wide, extends its disturbing influence over a comparable distance in all directions and deserves special treatment. Suffice it to mention at the present juncture that this complex includes in the south two large crater valleys, comparable to the Rheita Valley: the Bouvard Valley (it was thought to be a crater), some 200 km long and still technically on the Earthward side, and the narrower but longer trench, running 250 km or so, almost due north and south, along the 98th meridian.

Comparable surface rifts are to be found near a great thalassoid, centred on the 130th eastern meridian and the 70th southern parallel, beyond the Leibnitzes and the crater Amundsen as seen from the Earth, where, incidentally, the *Chart* appears to mis-construe the lie of the land, with its clearly orthogonal patterns, as ill-defined flowing lines.

Various crater alignments and concatenations, both linear and arcuate, occur among the vast uplands of the Moon's back, and the largest of them all is a great shallow trough running south with a slight westerly bias from its geometrical centre. It is over 1,000 km long and up to 50 km wide in places, ending up in the south in a peculiar crater formation, as though a giant footmark. This footmark pattern is repeated several times in this region, but nowhere else, with some variation in scale and detail, and attains gigantic proportions about the latitude of 40° S. It is a very inter-esting point that distinctive combinations and shapes of craters are uniquely associated with definite selenographical areas.[4]

Thus we have on the Moon's 'farside' linear and sublinear sur-face fractures, which may be open or sealed and run out into crater chains or crater alignments as, for instance, in the meridional trends north of Tsiolkovsky, but graben-rilles, typical of the Earthward hemisphere, especially in and near the maria, seem to be wholly lacking. An intensive search in the *Orbiter* photographs has revealed only one short sinuous rille; although there may be some small ones in Mare Australe and the adjacent walled plains. The relative scarcity of rayed craters cannot be a chance coincidence either. All these features seem to be associated with the maria.

With these reservations the 'farside' bears a close resemblance to the uplands of the Fourth and particularly the Third Quad-rant, whose natural continuation it indeed forms. The 295-km Bailly is a thalassoid, and Clavius, though smaller (230 km), is a comparable formation. Outlines of vast walled plains, of which Hörbiger alone is complete in the south, can be traced in a huge chain northwards along the east coast of Mare Nubium and Cognitum quite easily, and less clearly elsewhere in the Nubian and Imbrian plains.[2]

It would thus seem that the maria are a late selenological arrival, and the original surface structure of the 'nearside' and 'farside' was essentially the same – vast crater-lands with little or no lunabase. The fact that the maria are so heavily concentrated on the 'nearside' points to their connection with the Earth and the suspected capture of the Moon by it (p. 22).

REFERENCES

1　BARABASHOV, N. P., MIKHAILOV, A. A., and LIPSKIY, YU. N.,
　EDITORS (1961). *Atlas of the Other Side of the Moon.* Pergamon
　Press.

2　FIRSOFF, V. A. (1957). 'Palaeography of the Nubian and Imbrian
　Plains', *J.B.A.A.,* **67** (4).

3　FIRSOFF, V. A. (1959). *Strange World of the Moon.* Hutchinson,
　London.

4　FIRSOFF, V. A. (1961). *Surface of the Moon.* Hutchinson, London.

5　FIRSOFF, V. A. (1966). *The Moon.* The New American Library,
　New York and Toronto.

6　KOPAL, ZDENEK (1968). *Exploration of the Moon by Spacecraft.*
　Oliver & Boyd, Edinburgh and London.

7　NASA (1967, October). *Lunar Farside Chart (LFC–1),* 2nd Edn.

8　NASA (1966–67). *Lunar Orbiter Photo Index, I–V.*

9　PETTIT, EDISON (1940). *Astrophysical Journal,* **91**.

10　'The Colour of Space', *The Times,* 6th Jan. 1969.

11　WILKINS, H. P. (1954). *Our Moon.* Muller, London.

12　WILKINS, H. P., and MOORE, P. (1956). *The Moon.* Faber, London.

8

Lunar Tectonics

The geological processes that have shaped the surface of the Moon are basically the same as those operating on the Earth, but with a change of emphasis and a profound modification of effect due to the sixfold reduction in the force of gravity, with the multifarious consequences this entails. The matter has already received some attention in Chapter 5 : here we are concerned with tectonics.

The terrestrial, or telluric, rocks are generally compact and relatively heavy, averaging 2·7 in density on continental surfaces, which are thickly encased in sediments. These are derived mainly from light-coloured acid plutonic formations, typified by granite, and for the most part deposited in and under water. The dark-coloured basic rocks, which form the ocean beds and occur occasionally as a land overlay, for instance, in Iceland and Northern Ireland, are heavier still. The relief of the land is being continuously moulded by subaerial denudation, water and ice erosion, which quickly remove friable volcanics that may be comparable to the lunar surface formations.

On the present information we can no longer wholly discount similar action on the Moon, but this appears to have been episodic and largely localized. For the most part it may be assumed that the compacting agency of liquid water has been absent, and this in combination with low gravity will make lunar crustal rocks porous and generally lighter than water down to an unspecified depth, which may run into kilometres, especially in the terrae.

Rocks so textured will be stiff and brittle. Being also very light *per se* and owing to the low gravity, they will exert little pressure even at the base of a big mountain (p. 59). Under high pressures compact terrestrial rocks exhibit plastic behaviour and can even flow like a very viscous liquid (rheidity – see p. 16), but a low *confining pressure* inhibits folding and favours fracturing.[2] A descent of many kilometres towards the Moon's centre will be needed before confining pressures comparable to those of our upper crust are reached. This is why folding is virtually unknown on the Moon (with one or two possible exceptions, p. 162),[13] whilst fractures are numerous and deep-seated. A lunar surface fracture will thus generally penetrate much deeper than its terrestrial counterpart before being sealed, and so is correspondingly the more likely to tap a reservoir of magma, hot gases or liquids. This means that lunar fractures will be particularly prone to give rise to volcanic activity, pneumatolysis (=rock metamorphism by hot fluids) and mineralization.

Be this as it may, fractures and faults are familiar features of our geology. We have *dip-slip faults,* where the rocks have moved up or down along the fracture. If this movement, measured vertically by the *throw*, has been down the slope, or *dip* (dip is measured from the horizontal plane, *hade* from the vertical in the direction of movement), of the fault, the fault is *normal*. In a *reverse*, or *thrust,* fault the rocks have risen along the dip (against the hade), and if the dip is gentle, as it will be in poorly coherent, or *incompetent* rocks, and the movement continues, an *overthrust* may develop. Overthrusts are known in the Alps and the Scottish Highlands where the strata from the upper side of the fault have been pushed for tens of miles over its lower side. Most terrestrial overthrusts originate in folding, but this is a secondary feature, as the fold must be faulted before an overthrust is produced.[2, 17]

Another type of movement is found in the *strike-slip* (also known as *tear, shear, wrench* or *transcurrent*) *fault,* where the two sides of the fracture are displaced horizontally relatively to each other as exemplified in San Andreas Fault in California, responsible for the disastrous earthquake which destroyed San Francisco in 1906. Horizontal offsets in the Great Glen Fault in Scotland attain 60 miles. If, as we look along the fracture, its right side appears to have moved away from us the fault is *right-handed,* or *dextral*; if the other way, it is *left-handed,* or *sinistral*.

In most faults there is some movement both along and across

the fracture, and there may be various further complications, such as *step-faulting*, with several parallel faults, which may differ in dip and throw and result in a terraced configuration. A special development of lunar interest is a *concertina fold*, which does not involve true folding, but merely a large number of oppositely inclined step-faults on the two sides of a ridge. Something of this nature appears in one of the central peaks of Copernicus, and in the walls of some craters which display a multitude of parallel fractures. There is no clear distinction between a fault, a fracture and a rock joint : it is merely a matter of degree, or of the intensity of the stress applied to the rocks. *Horsts* and *grabens* have already been described on p. 74 f. Some of the lunar grabens appear to be bounded by strike-slip faults of opposite transcurrence.

On the Earth the rock faces grinding against each other along a faultline often become smoothed, and are known as *slickensides*. Alternatively they may be crushed, the gap between the two sides of the fault being filled with *fault breccia*, which, if the brittle nature of lunar rocks is borne in mind, seems to be the more likely on the Moon. As already mentioned, a fracture may be injected with magma, and if this reaches the surface the result will be a fissure eruption, which are of frequent occurrence in Iceland. Volcanic vents may open up along such a magmatically active fracture, giving rise to a crater chain, a development also well known from Iceland and the slopes of many volcanoes, such as Etna. Volcanoes often stand at intersections of such lines of tectonic weakness. A fracture filled with solidified magma is known as a *dyke*; but magma is only a special case of incompetent rock, and shale, sandstone, and more particularly salt and ice may also be forced up along a fracture line and produce dykes.[17] It must, however, be remembered that the upward impelling force is due to a pressure differential, which is a function of gravity and heat, and, other things being equal, it will be much lower on the Moon, so that magma will not generally be expelled from a fracture to the surface as vigorously as it is in terrestrial conditions.

Faults may be straight or arcuate, although strike-slip faults must be more or less straight by their very nature. There are also *ring-faults* defining cauldron subsidences, which will be considered in more detail in the next chapter.

Crustal stresses vary, both locally and in time, and the same fracture will respond differently to the changing stress. It may then

be spoken of as a *tectonic lineament*. Vertical stresses are usually
due to differential crustal loading and are related to isostasy,
although they may also arise from body tides, intrusions or with-
drawals of magma or other supporting medium below the ground,
and similar developments. A vertical stress applied at one point
will pull the crust up or down, and this in turn must generate
horizontal stress farther afield. Subsidence will generally give rise
to a *compressive,* uplift to a *tensional* stress, attendant on crustal
shortening and dilatation respectively. Such effects may be local,
regional or global. If we exclude folding, compressive stress is
relieved by upthrust, overthrust, extrusion of yielding materials,
such as magma, horst-faulting, uptilting of crustal blocks and so
on. Conversely, tensional stress will cause the existing fractures to
widen and new ones to open, the ground subsiding between them
into grabens and rift valleys or basins and cauldrons. It may be
accompanied by the downtilting, horizontal shifting and rotation
of crustal blocks. But any relief of stress will cause its direction to
be reversed, setting up a kind of wave pattern.

This is beautifully illustrated by the case of Mare Orientale.
There is a central area of subsidence, generating a compressive
stress, surrounded by a ring of tension and fracturing of the surface, which again is followed by a ring of compression and uplift,
itself surrounded by an area of collapse, ringed by uplift and
extrusion, the sequence being repeated once more with a widening
of the stages and a decrease in the amplitude of the disturbance.
It is found that in tectonic subsidences the radii of the successive
ring-scarps increase in the ratio of $\sqrt{2}:1$. In the case of Mare
Orientale the progression is approximately $4:6:8.5$, which is close
enough to theory, especially if it is considered that the innermost
scarp is very indistinct, much broken up by oblique lineaments,
and may comprise two separate elements. Mare Nectaris is the only
other formation where a somewhat comparable structure can be
discerned, and Fielder finds it to conform to the same pattern.[8]

The existence of overthrusts is not so easy to determine by
telescopic observation or even orbital photography. They are
favoured by stratification, and it is doubtful if this is well developed
on the Moon, although superimposition of volcanic deposits (pp.
57 ff.) may result in layers differing in texture and coherence. A
small piece of rock picked up by *Surveyor III* (p. 58) is said to
have been laminated, which represents small-scale stratification.

The regularity of terracing on the inner glacis of some craters, such as Hell, suggests excavation in stratified rocks,[13] but may be misleading.

In any event the reality of lunar overthrusts is often denied. To what extent this may be due to the tendency to bend the data into conformity with the exogenic – impact – interpretation of the lunar surface, is another matter.

However, in the central highlands of the Earthward hemisphere there are clear signs of a thrust directed about 10° east of the meridian, when looking south. This thrust defines the southern coastline of Mare Nubium, and is very strongly marked between Hörbiger and Werner, where the walled plain of Purbach appears to have been pushed bodily over Regiomontanus through about 20 km. That this is not a matter of simple overlap is shown by the bounding fracture line extending into the neighbouring formations, where a small crater near Blanchinus is characteristically deformed by it. Blanchinus itself is somewhat pear-shaped, its wider north-western part having a differently structured and differently coloured floor. In fact, it appears to have arisen through coalescence of two rings, the south-eastern one having been completely destroyed on the side of the overthrust. The situation is similar farther east along the line, although Playfair is unaffected and clearly posterior to this land movement. Beyond it, however, Playfair A is again deformed.

I have given some further examples of possible overthrusts in this area in a paper, 'On the Tectonic Grids of the Moon', published in the *Journal of the British Astronomical Association* in 1957,[12] but not much notice appears to have been taken of this in later work.[5]

However, dip- and strike-slip faults are beyond dispute. The Straight Wall is a conspicuous example of dip-slip faulting and has already been described on p. 74. The west wall of Albategnius is cut by a strike-slip fault with an offset of about 3 km.[15] Larger offsets abound. The uplands in this region are criss-crossed by an intricate lattice of fractures, and so, too, are the surroundings of Mare Orientale, where some complicated displacements of large land blocks have occurred (Plate 5).

At this juncture we may profitably direct our attention at the Caucasus and the Alps on the north-eastern borders of Mare Imbrium.

These ranges are clearly part of the mountain girdle round the Mare, as are the Apennines farther south (in the west the girdle has been more or less completely destroyed and the Imbrian Plain passes imperceptibly into Oceanus Procellarum). Whereas, though, the Apennines, if not undisturbed by subradial and other shifts, form a substantially continuous scarp towards the Mare, and remain aligned with the Caucasus after a 100-km gap, the Caucasus itself is shattered into distinct blocks by well-defined fault-lines and is at least 200 km out of line with the Alpine front, which is approximately maintained by the craters Cassini and Theaetetus instead (Plate 2).

Not only the fronts, but the hinterlands of the Caucasus and the Alps match and can be brought into near-coincidence by rotating the former through 60° westwards about its southern tip. There is, indeed, multiple evidence of land creep away from Mare Imbrium in this area. This movement may have been associated with the formation of Mare Serenitatis and obviously occurred after Mare Imbrium and its mountain girdle had been fully developed. It is one of surface dilatation, the mountains being split by deep grabens and rift valleys, bounded by strike-slip faults.

The situation is classically illustrated by the Alpine Valley (p. 71). 120 km long and up to 16 km wide, it shows a sinistral offset of 32 km (Fielder):[8] the two sides can be brought into approximate coincidence if the south-west bank is advanced towards the Mare by this amount. The offsets, however, are not the same for all features, and Fielder gives 20 km as the shift needed to restore the continuity of a prominent horst lineament near the southern end of the Valley. This points to a gradual growth of the Valley, the horst having been formed after its initiation. There is also a dextral (clockwise) rotation through some 60°, which he does not mention, bringing the ridge into approximate alignment with the Caucasus front, which has experienced an opposite angular displacement by the same amount.

The Caucasus is split by still wider rifts, some 30 and 40 km across in the neighbourhood of Aristoteles, while the Apennines show dextral offsets of up to 40 km subradially to Mare Imbrium. A trapezoidal land block in the southern half of the Caucasus is worthy of special notice. It is 90 km deep and some 65 km wide on the Imbrium side, and has been shifted as a whole, with some shattering at the south-west end, by 16 km in the north and 10·5 km in the south, which has twisted it out of position by about 20°.

It is interesting to note here that the stress increased northwards, which fits the pattern of the off-meridional paired craters, such as Eudoxus-Aristoteles, Aristillus-Autolycus, Sheepshanks-Mayer and Protogoras-Archytes, in this area, the northern crater being always the larger of the two. The marial basins, too, show a similar trend. Beyond the equator, however, as we approach the south pole, the situation appears to be reversed, for instance, Aliacensis being larger than Werner.

Whatever the interpretation of the latter peculiarity, such land movements necessitate the existence, at least in the past and at least in the region of the maria, of a yielding substratum, capable of flow – probably rheid flow – and entraining the rigid and brittle surface formations. Movements of isostatic adjustment must, therefore, likewise have taken place.

Yet dislocations of this magnitude are confined to the maria and the adjacent *perimarial* terrain. In the lunarite uplands remote from the maria large strike-slip faults and rifts are less frequent, save for the circumpolar and librational zones.

Fielder[8] traces the Purbach-Regiomontanus fault over 270 km, which I think to be an underestimate, as the effects of this lineament are visible over at least a 1,000 km, and gives 26 km as the lateral shift, which I am unable either to confirm or deny. But, as already stated (p. 93), I regard this fault primarily as an overthrust.

In any case it is one of the many similar lines of crustal weakness, along which land movements and/or magmatic activity have taken place in a chronological succession difficult to determine, except sometimes in relation to ring mountains, which are themselves of uncertain age. A group of such approximately parallel or otherwise visibly associated surface fractures constitutes a *family* of lineaments. Some families are purely local. For instance, the faults and graben-rilles following the southern contour of Mare Humorum may be described as such a local family. There are also radial or subradial fracture families, one of which is associated with the large ruined, or undeveloped (see p. 74), ring of Letronne. A radial pattern of this kind may arise through an underground igneous intrusion, as in the star systems of dykes of Mull and Arran in Scotland, but it could also be due to meteoritic impact, although terrestrial geology knows no examples of this.

As we survey the Moon, however, especially in the Third and Fourth Quadrant, we cannot help noticing numerous alignments of such features as fractures, faults, rilles, crater chains, ridges, walls of large ring mountains and bulwark plains, affecting the general lie of the land in repetitive patterns. These are the lunar *tectonic grids*.

A single line or concatenation of such features is referred to as a *septum* (pl. *septa*), and represents a line of deep-seated crustal weakness (fracture). Crustal stresses will differ from place to place and epoch to epoch; some may be in sympathy with one, others with another component of the grid system, which will be differentially emphasized accordingly. The type of response will also vary : it may now be a strike-slip, then a dip-slip, a widening of the gaps with graben formation, or contrariwise a squeezing movement with the elevation of horsts, volcanic activity, or extrusion of magma, forming ridges, known as *tholoids,* and tumuli. Igneous activity will be considered in more detail in the next chapter, but it is apposite to repeat here that low surface loading must generally reduce the upward impetus of ascending magma, which may fail to reach ground level even along an open fissure. Moreover, the low gravity and barometric pressure will cause the exposed surface of magma (lava) to froth up violently, the expansion of the occluded gases having a chilling effect, so that a thick skin will form over it, once more impeding the advance. These are the reasons why tholoids are so common a feature of lunar landscape, and why fractures and crater chains so often pass into ridges.

Fracture grids are not peculiar to the Moon; they exist on the Earth as well, as has been pointed out by Hobbs, Umbgrove, Vening Meinesz and others. *Hobbs says* : [16]

'The recognition within the fracture complex of the earth's outer shell of a unique and relatively simple pattern, common to at least a large portion of the surface, obscured though it may be in local districts through the superimposition of more or less disorderly fracture complexes, must be regarded as of the most fundamental importance.'

L. U. de Sitter (1956) also affirms that 'There evidently exist in the earth's crust long fundamental cracks which are characterized by frequent movements along their contiguous faces, and by the fact that they penetrate the whole solid crust and occasionally constitute a way of access to the surface for basic and granitic magma'.[2] There are, however, good reasons why such grids must be even more important in lunar conditions. As already noted at the beginning of this chapter, the Moon is virtually immune to folding, which leaves no alternative to fracture as a response to crustal stress; owing to the nature of the rocks, these fractures will extend much deeper into the interior of the globe; this will be all the more so if the conception of crustal honeycomb proposed in Chapter 5 is accepted; the grids are most probably associated with the capture of the Moon by the Earth, involving a close approach, slowing-down of rotation and powerful body tides, which would make for the universality of the grid system. Indeed, we have already seen in Chapter 7 that the 'farside' lineaments are closely comparable to those of the Earthward hemisphere.

In any event there can be no question about the great importance of these tectonic grids for the development of the Moon's surface features. They are truthfully reproduced in small-scale structure, such as rock jointing, often giving the country a peculiar tartan-like appearance, referred to as *lattice*. It is not unusual in geology that the large-scale lie of the land is mimicked in miniature in features of and below a metre in diameter. This is certainly true of the Moon.

It is not easy to say who originated the idea of lunar grids. In Germany Kurd von Bülow spoke of *Gitter* (grids) in 1957,[1] but referred to similar ideas published by the Austrian geologist E. Suess in 1909. In Czechoslovakia the grid structure in the area of Sinus Iridum attracted the attention of O. Matousek, who published two papers about it in 1924.[5] But the greatest single contribution to the study of these features was made between 1944 and 1949 by the American geologist J. E. Spurr.[20-22] 1949 was also the year in which A. V. Habakov in Russia drew his map of lunar grids.[1, 5]

I knew only Spurr's work when I first tackled this subject in the 1950s, and my first paper on it was published in 1956.[11] In this I was soon to be followed by Gilbert Fielder. We, too, compiled, quite independently from each other or anybody else, provisional charts of lunar grids, which appeared simultaneously in 1957.[4, 12] Mine was revised and republished two years later (Plate

17), and in its final form in 1961,[14] in which it seems to remain to this day the most comprehensive study of the tectonic lineaments on the Earthward side of the Moon, though, no doubt, it, too, can be improved upon.

As Fielder says in *Structure of the Moon's Surface*[5] (1961) (p. 185) 'it is most gratifying to note that there is a substantial amount of agreement between the different observers', although they had no knowledge of one another's work. This is perhaps not very surprising, as most of the grid septa are quite unmistakable. He goes on to say that his own map comes closest to von Bülow's[1] – so does mine for that matter, perhaps even more so.

There was, however, a certain difference in approach between us. Fielder had a tendency to schematize the situation and to let his conception of the situation be dominated by a 'system radial to Mare Imbrium',[3] the latter of which applies *a fortiori* to Habakov, who shows substantially one system of lineaments only, except for the fault-lines at the foot of the Apennines.

That Mare Imbrium is associated with a system of septa that is roughly radial, or subradial, to it, cannot be reasonably denied. But this system itself can often be broken up into families of lineaments which are 'subparallel' between themselves as well as being intersected by other systems of septa at various angles. Its prominence is at least partly spurious, especially in the Ptolemaean highlands, and due quite simply to the fact that these lineaments run nearly north and south and so cast clear shadows under both the morning and evening Sun, whereas those septa which approximate to a latitudinal distribution stay shadowless and are difficult to see. This point did not escape von Bülow's[1] notice either, and I have made a special effort to eliminate the bias due to lighting by the study of photographs taken at different phases and telescopic observations under every possible angle of illumination. For this reason my map differs from others mainly in the inclusion of numerous septa that form small angles with the direction of sunlight (Plate 8a).

Fielder has done much praiseworthy work on lunar grids, but always with an eye on a mathematical solution. This presents its advantages, but also involves a certain risk of losing touch with reality. Like von Bülow and myself, he speaks of a 'triple grid system', but we do not seem to mean quite the same thing by it. Indeed, in my interpretation each grid has two components approximately orthogonal to each other, which makes six dominant trends in all. Fielder, however, distinguishes only three trends: A, B and C, the first two being oblique and the last meridional. He also makes a slurred reference to the existence of lineaments at right angles to C, but is mainly preoccupied with A and B. In his *Lunar Geology*[8] (1965) (p. 120) he sums up the situation thus:

(i) They (the grids) cross the face of the Moon in closely defined directions for thousands of kilometres.

(ii) The poles of the axes of symmetry of ... A and ... B are located at position angles (measured from the north pole of the Moon through east) of 135° and 45°, respectively.

(iii) The angle of intersection of ... A and B varies systematically, but the Moon's equator and centre meridian are axes of symmetry.

(iv) Both systems were produced over the same, long period of time ...'

He, further, finds (v) 'direct evidence for predominantly dextral offsets in A and of predominantly sinistral in B'. Measurements made in *The Rectified Lunar Atlas* indicate that the average angle between the A and B lines is 88°, although it may occasionally exceed 90°.

From this analysis he infers that the origin of the grids must be sought in a global stress, and not in any random action, such as meteoritic bombardment; that this stress must be related to the Earth; that conditions (iv) and (v) 'may ... be satisfied if the strike-slip faults of the systems A and B were produced by a meridianal pressure in the Moon's crust' (p. 121). According to Mohr, Anderson and others this kind of compression would generate fractures along the main stress and intersecting patterns of shear fractures (strike-slip faults), arranged in pairs of opposite shift, on their two sides. Fielder[8, 9] draws a barrel-shaped pattern of the 'A and B grids', referred to the central meridian of the Earthward hemisphere and the equator as axes of symmetry (condition [iii]), and assimilates it to the four-cell convection system in the mantle of the Moon proposed by S. K. Runcorn.[19] Fielder shows that the angles between A and B are always bisected by the maximal stress due to the convection.

This is an interesting and in many ways fruitful line of thought, but it encounters various difficulties, the foremost of which is that the same system of fractures extends to the other side of the Moon without any change of angular relationship, which shows that any symmetry about the central meridian on the Earthward hemisphere must be a purely secondary effect. Further, there exists another system of paired oblique grids, as will be clearly seen from the *Orbiter IV* photograph of the Ptolemaean highlands reproduced here (Plate 8a).

One oblique trend is represented by the crater chain at the north-east wall of Ptolemaeus and the parallel lineaments defining the opposite walls of this and the neighbouring walled plains; while the coresponding 'vertical component' is marked out by the crater chain striking from Thebit past Hell towards Tycho. The second system makes an angle of 30 – 35° with the two components of the first, and the third, 'off-meridianal system' deviates by 60° or so westwards from the first as seen from the north. This is the trend defined by the Straight Wall and the ridge lineaments in the Maria, Nubium and Cognitum. In the Ptolemaean highlands it may be confused with the second system, but the point is that both trends occur together now and again as distinct entities, for instance, south of Lalande. The 'sublatitudinal' components of this third system are difficult to see, except in graben-rilles, owing to parallel illumination, but one master fracture can be traced right across the Mare and Oceanus Procellarum to the western limb.

The same fracture grids reappear around Mare Orientale.

To sum up, it seems that the explanation, proposed in *Surface of the Moon*[13] (1961), that the grids arose through the tidal action of the Earth on a closer asynchronously rotating Moon, gives a better account of the facts. This would involve alternate compression and dilatation, generating septa, roughly at right angles to each other, and in three systems, substantially as described. Nor is any different mechanism required for the 'farside', only that the formation of the maria on this side of the Moon must be associated with the phase in the development of the lunar orbit, when the Moon, already in slow rotation, reached its point of closest approach to the Earth, and possibly entered Roche's Limit (Kopal,[18] see pp. 21–23).

This would have happened between 1·5 and 2 aeons ago, in pre-Cambrian times. The reciprocal grids of the Earth may date from the same period.

The slowing-down of the Moon's axial spin would have resulted in at least a partial adjustment to the figure of hydrostatic equilibrium, involving a decrease in polar flattening and the collapse of the equatorial bulge. The first would lead to tensional stress in the polar regions, which fits in well with the existence there of rift-valleys, high mountains, deep craters and large thalassoids, especially in the south. In the northern circumpolar zone part of the stress would seem to have been absorbed in the formation of the maria, but the whole area north of the Mare Frigoris appears to have undergone an intense metamorphism, the walls of the old rings having slumped and their floors been levelled, everything being thickly peppered with craterlets, some of them in rows. These craterlets are all of approximately the same size, and the effect has no counterpart anywhere else on the Moon. They are invisible telescopically, and have only been revealed by orbital photography. But they seem to bear out Spurr's[21] original idea that this part of the Moon was reheated, consequent on the formation of the maria, until the rocks became plastic, the craterlets having been produced by the escape of gases occluded in them.

In the equatorial zone the sinking of the rotational bulge would result in east-to-west crustal shortening, which Spurr called *appression*.[22] This clearly manifests itself on both sides of the Moon in a frequent breaching of the walls of old rings in the north and south, and the coincident development of the so-called *deltoids*, or the *Gothic-arch effect*, on the outside glacis.[9, 10] This is beautifully exemplified by the thalassoid halfway between Mare Orientale and Mare Moscoviense on the 'farside' (Plate 6).

At a later stage the maria appear to have behaved like rigid shields against which the contracting brittle crust was buckled, piled up and upraised.[13] Thus both the sinking of the marial basins and the elevation of their girdling mountain ranges would be a comparatively protracted process, perhaps continuing to this day, initiated by the tidal stresses of the Moon's capture and the equatorial adjustment to the reduced rate of axial spin. A concomitant shift in the position of the polar axis may be suspected, the original marial belt having been more nearly coincident with the equator.

Yet while the Jura Mountains and the whole of the Iridum highland were raised and tilted more or less *en bloc*, the stresses farther east, and especially in the Caucasus, were much more severe and complicated, with shattering, wrench-faulting and volcanic activity along the marginal fractures.

Our alpine mountains are produced by the squeezing-out of marine sediments from geosynclinal troughs, and in the absence of seas this process cannot operate on the Moon, or – to put it more cautiously – not on its surface. As already mentioned in Chapter 6, such linear mountains as the Lunar Alps are primarily tholoids, extruded along magmatically active surface fractures, and many of the peaks are obvious extinct volcanoes with blocked summit craters. They stand high above the hinterland, and so are not just the edge of a tilted land block. This is true even of the Apennines, where the block structure is unmistakable, as the edge of the scarp above the Imbrian plain is much higher than the immediate background and includes some well-individualized volcanic peaks.

We will return to this problem in another context, but the occasion invites comparison with the 5,000-foot Altai scarp, which is clearly similar to the Apennines, but very ancient and subdued by long ages of erosion. This, incidentally, goes to show that the maria are not even approximately contemporaneous, although those of the Imbrian system probably are.

Fielder[8] follows Spurr[20] in suggesting that in the Altai the 'selenofault' curves round the rim craters, instead of cutting through them in the manner of other deep-seated fractures. I think, however, that this distinction is spurious, the salients in the Altai scarp having been produced by strike-slipping, and the craters being the stumps of the volcanoes that burst out along the main fracture, just as they did in the Caucasus or the Apennines. The difference of appearance is due solely to age.

REFERENCES

1 BÜLOW, KURD VON (1957). 'Tektonische Analyse der Mondrinde'. *Geologie*, **6** (6/7), p. 565 ff.

2 DE SITTER, L. U. (1956). *Structural Geology*. MacGraw Hill, London.

3 FIELDER, GILBERT (1955). 'A Study of the Valley System Radial to Mare Imbrium', *J.B.A.A.*, **66** (1), p. 26 ff.

4 FIELDER, GILBERT (1957). 'The Lunar Grid System', *J.B.A.A.*, **67** (8/2), p. 314 ff.

5 FIELDER, GILBERT (1961). *Structure of the Moon's Surface*. Pergamon Press, Oxford.

6 FIELDER, GILBERT (1962). *The Observatory,* **82** (930), p. 196 ff.

7 FIELDER, GILBERT (1963). *Geoph. Jrnl. R.A.S.,* **8** (2), p. 187 ff.

8 FIELDER, GILBERT (1965). *Lunar Geology.* Lutterworth Press, London.

9 FIELDER, GILBERT (1966). 'Convection in the Moon: A Boundary Condition', *Geoph. Jrnl. R.A.S.,* **10**, p. 437 ff.

10 FIELDER, G., and WARNER, B. (1962). 'Stress systems in the vicinity of lunar craters', *Planetary and Space Science,* **9**, p. 11 ff.

11 FIRSOFF, V. A. (1956). 'On the Structure and Origin of Lunar Surface Features', *J.B.A.A.,* **66** (8), p. 314 ff.

12 FIRSOFF, V. A. (1957). 'On the Tectonic Grids of the Moon', *J.B.A.A.,* **67** (8/2), p. 309 ff.

13 FIRSOFF, V. A. (1961). *Surface of the Moon.* Hutchinson, London.

14 FIRSOFF, V. A. (1961). *Moon Atlas.* Hutchinson, London.

15 FIRSOFF, V. A. (1967). 'Preliminary Selenological Analysis of *Ranger 9* Photographs', *J.B.A.A.,* **77** (2), p. 106 ff.

16 HOBBS, W. H. (1911). 'Repeating Patterns in the Relief and the Structure of the Land', *Bulletin of the Geological Society of America,* **22**, p. 163 ff.

17 HOLMES, ARTHUR (1965). *Principles of Physical Geology.* Nelson, Edinburgh and London.

18 KOPAL, ZDENEK (1966). *Nature,* **210**, p. 1,347.

19 RUNCORN, S. K. (1967). 'Convection in the Moon and the existence of a lunar core', *Proc. of the R.S., Series 4,* **296** (1446), p. 270 ff.

20 SPURR, J. E. (1944). *Geology Applied to Selenology – The Imbrian Region of the Moon.* Science Press, Lancaster, Pa.

21 SPURR, J. E. (1948). *Geology Applied to Selenology – III – Lunar Catastrophic Theory.* Rumford Press, Concord.

22 SPURR, J. E. (1949). *Geology Applied to Selenology – IV – The Shrunken Moon.* Business Press.

Abbreviations used

J.B.A.A.	=Journal of the British Astronomical Association.
Geoph. Jrnl. R.A.S.	=Geophysical Journal of the Royal Astronomical Society.
Proc. R.S.	=Proceedings of the Royal Society.

9

What are the craters?

The Moon is full of holes, big and small, like Swiss cheese, and, as indicated in Chapter 5, this frivolous analogy may come nearer the truth than learned verbiage laden with long Greek and Latin words, for the subsurface layers of the lunar globe seem indeed to resemble Swiss cheese in structure. On the surface, however, the holes are 'sealed' and appear as comparatively shallow depressions, round, polygonal and at times more or less irregular, walled or unwalled on the outside. The last-mentioned are usually small and referred to as crater-pits, but in the larger examples an upraised rim is invariably present, becoming more and more complicated with increasing size, until the term 'crater' no longer conveys the essence of the structure that would be more accurately described as a 'mountain ring'. Nevertheless, these walled depressions are generally known as craters, although they may be differentiated as ring mountains, crater plains, bulwark or walled plains, and finally thalassoids (p. 82), which become maria if dark-floored, in the order of ascending scale. K. von Bülow, Jack Green and G. C. Amstutz[4] have proposed a division into four classes, with further sub-classes, which presents certain advantages, but until it has become firmly established tends only to make the confusion worse confounded.

In any case a general morphological description of these features has already been given in Chapters 6 and 7, and there is no point in going over all this once more. But the disparity in the size and shape of craters is so great that it is by no means certain

Plate 7. Lunar tectonic grids. This chart by the author was originally published in *Surface of the Moon* (Hutchinson, 1961). South is at the top, as in Plate 1, to which please refer for comparison. .... denotes tight crater chains. Interrupted lines indicate suspected grid lineaments. (See also Plate 8a.)

Plate 8a. Grid patterns in the Ptolemaean highlands. Ptolemaeus is the large
walled plain near the middle, Alphonsus
immediately below it. A portion of
Orbiter frame IV–102M$_1$.

Plate 8b. A portion of the marginal heights of Flamsteed
P, showing a 'welt', tors and boulders along the edge
the scarp (III–199H$_3$). *N.A.S.A.*

Plate 9. A portion of III–121H₂, showing the lava-flooded ring south of Tsiolkovsky (see also Plate 14).

10a. The crater is about 500 ft. (say 165 metres) across, and the unlikely official explanation of its double walls is landslide after meteoritic impact (p. 137). This is a portion of an *Orbiter-III* frame 204-H.

10b. The second crater is about twice as large and shows several secondary explosion centres within it. Many subdued craterlets indicate a similar original structure, which cannot be due to impact and is most probably the result of repeated phreatic eruption (p. 138). HRII–177₂

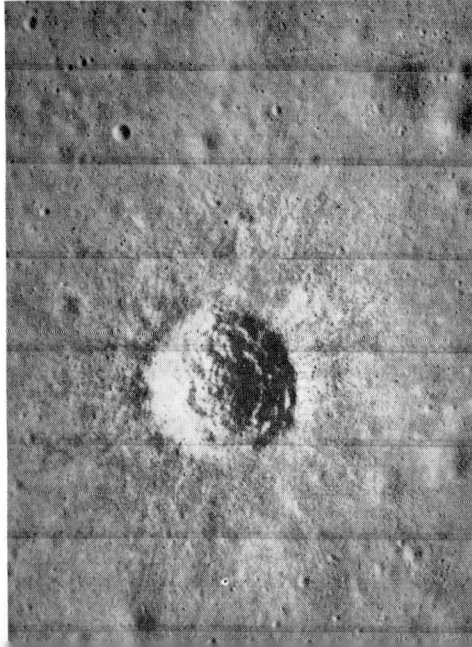

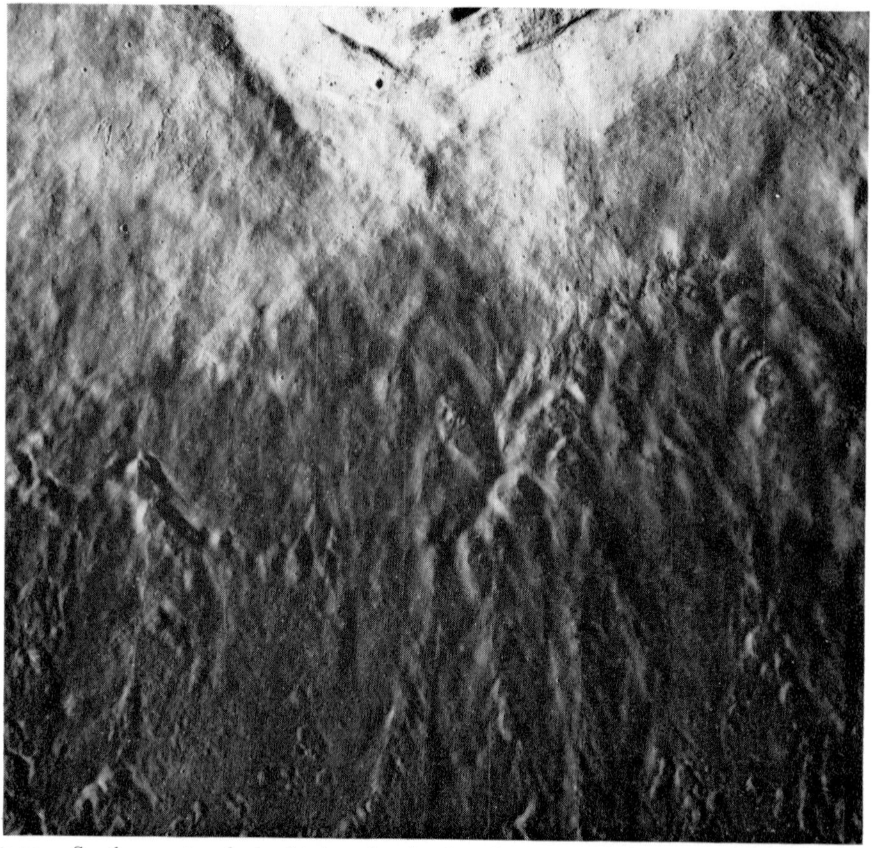

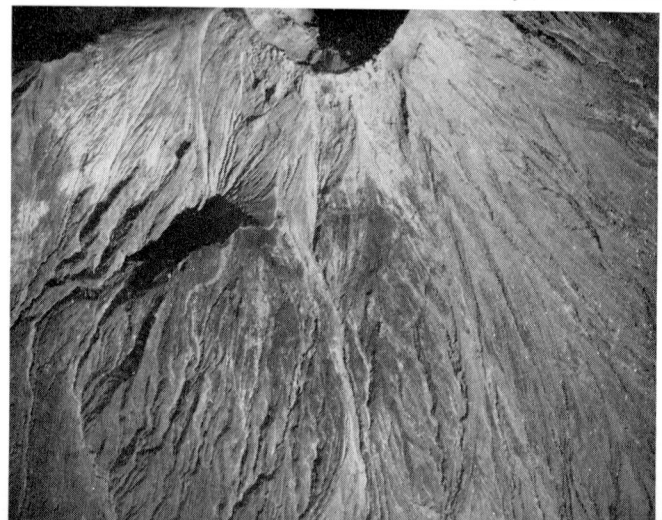

Plate 11.   Southern outer glacis of Aristarchus (11a), to be compared with the slopes of Oldonyo Lengai in 11b, bearing in mind that the latter has been scoured by torrential tropical downpours, whereas water erosion of the Aristarchus glacis has been at most slight. Oldonyo Lengai is an active volcano in Tanzania.   *a.* (V–196M.)   *N.A.S.A.*

b. A photograph by the Royal Air Force. *Department of Lands and Surveys, Dar-es-Salaam* that they have all been the work of the same forces; nor must it be overlooked that very different processes may produce superficially similar forms.

In fact, the nature of these processes has been the subject of an age-long and embittered scientific controversy that has not run its course yet, despite all the new evidence of the close-up photography of the lunar surface, neither the protagonists of the meteoritic, or exogenic (external), nor of the plutonic, or endogenic, explanation being prepared to yield ground. Yet the high resolution of the probe photographs makes much of the earlier telescopic study and selenological thinking based on it obsolete, with the only reservation that, in order fully to assess the nature of lunar terrain, it must be seen under different angles of illumination, which these photographs seldom afford.

It may, further, be justly observed, as Moore and Cattermole do in *The Craters of the Moon* (1967),[15] that some ideas, such as the differential contraction or expansion of the crust and the interior, proposed by I. Ruud and A. Fillias respectively, 'cannot be dismissed out of hand' (p. 21). Indeed, even patently preposterous suggestions, to take Voigt (1904) and Beard's (1917) notion that the craters of the Moon are ancient coral atolls, Ocampo's stern warning to mankind (1949) that they are scars of an atomic war that has wiped out a lunar civilization, or Fountain's (1925) more reasoned argument that, as the Moon lost its atmosphere, its oceans froze under a coat of dust, the ice melting here and there to form craters, may occasionally point a useful line of thought, if only tangential to the original conception. A brief historical review is, therefore, not out of place.

Mention has already been made (p. 3) of Robert Hooke's experiments in the seventeenth century with boiling alabaster and his conclusion that the craters were the solidified rims of bursting bubbles on a hardening surface of the Moon convulsed by similar ebullition. This idea was still considered seriously by E. de Beaumont in 1831 and some others, but is obsolete in its original form. Nevertheless, we have seen in Chapter 5 and recalled at the outset of this one that the crustal honeycomb of the Moon may contain large cavities filled with volatiles, i.e. 'bubbles', and these may trigger off saucer-like subsidences on the surface, giving rise to craters and large walled plains in particular.

Most early lunar observers, including Schroeter and the Herschels, however, interpreted the craters as volcanoes pure and simple. They were not greatly perturbed by the thought that the largest known volcanic vent had a diameter of about 4 km, whereas the lunar walled plains could be nearly 300 km wide (Bailly), and that their floors were usually (if not always) depressed below the level of the surrounding country, which was quite the reverse of the terrestrial situation.

Chacornac's approach (1864) was more sophisticated: he thought that the mountain rings had been produced by repeated violent explosions, and drew attention to the differences of age between various formations, as well as to their central peaks.[15] It is the latter that caught the fancy of the British selenographers, Nasmyth and Carpenter, and in 1874[16] inspired the hypothesis of 'central fountain'. In this the central peak, whose summit is often cratered, is thought of as the actual volcano, while the walls around it are supposed to have been built up from the loose material (nowadays known as *tephra*) emitted by it. No comparable structures are known on the Earth; the hypothesis fails to account for the depressed floor and for the inward scarps of the ring. But it is not wholly mistaken inasmuch as the volcanic nature of central peaks is not seriously disputed, and they may have emitted showers of ash, lapilli and dust.

J. B. Hannay (1892), H. Faye (1896) and N. Boneff (1936) sought to eschew these difficulties by recourse to tidal action and fissure eruption. They held that the Moon was at one time much closer to the Earth than now and in slow asynchronous rotation. The tides drew magma from the interior alternately towards and away from the surface, in the manner of a pump, and some of this magma repeatedly oozed in and out of the fractures partially solidifying and building up mountains and crater walls.

This idea of 'marginal effusion' is not without merit, and we shall have occasion to return to it at a later stage. It was also favoured by W. H. Pickering[18] in the U.S.A. and P. Puiseux and H. Loewy in France. The main point, however, made by these selenographers and H. G. Tomkins in England was that the analogy between the lunar craters and the terrestrial volcanic vents was a false one, and that the former must be properly compared to *calderas of collapse*, commonly produced by the withdrawal of magmatic support from beneath a volcanic cone or a *blind puy* (i.e. a dome raised by a subterranean magmatic intrusion which did not gain access to the surface and opened a vent) though ocasionally by an explosion, the volcano blowing off its top, as did Krakatoa in 1883.

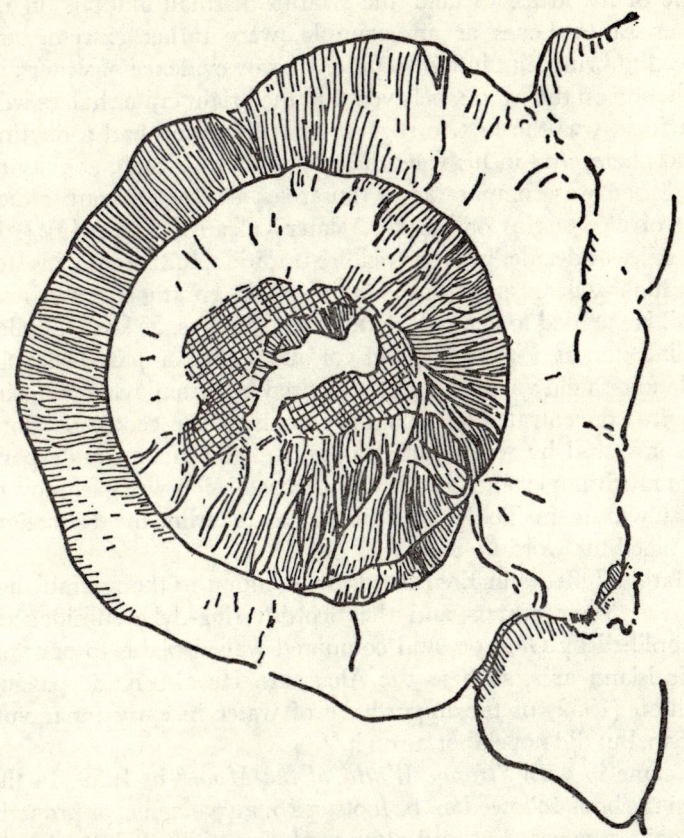

Fig. 4. The volcanic ring of Hverfjall, Iceland, with a central peak, drawn from an aerial photograph.

In his book, *The Moon*,[18] published in 1903, Pickering gave an impressive list of volcanic calderas, including the 15-mile Lake Bombon in the Philippines and the Bezimianniy caldera in Kamchatka, as well as the Japanese Volcano Bay, 25 miles in diameter. These were structures comparable to the lunar ring mountains, if only of medium size; but, it was argued, the reduction in gravity on the Moon, by making the overlying strata easier to lift, would favour magmatic intrusion, as well as increase the effect of volcanic forces, thus bridging the remaining gap.

Pickering was a good observer and an original thinker, although some of his ideas, to take 'the swarms of small animals' in the crater Eratosthenes as an example, were rather extreme and served to bring him into disrepute. He saw evidence of water and ice action on the Moon, believed that the bright crater haloes were hoarfrost, was the first to suggest that the Moon had a daytime atmosphere, and in his view the present-day volcanic activity on the Moon was comparable to fumaroles and geysers rather than that of a Vesuvius or Hekla. A later vulcanist (1944–49) J. E. Spurr[21] was decidedly in the hellfire tradition, but he, too, insisted that lunar vulcanism consisted mainly of an 'enormous gas exhalation'. He tended to view such craters as Aristillus, 55 km (35 miles) in diameter, as a special kind of volcano, to be compared with the Icelandic shield volcanoes, characterized by summit calderas with activity concentrated in secondary vents at the centre or along the rims, and by wide gentle slopes. But the lunar counterparts were much larger and the bulk of their emission was gas, most of which was steam. To describe this and to underline the distinction, he coined the word *lunavo*.

Harold Jeffreys in England drew attention to the resemblance between lunar craters and the Scottish ring-dyke subsidences, exemplified by Glencoe, and compared walled plains to our volcanic island arcs, such as the Aleutians. He also made passing mention (1952) of the importance of water in early lunar vulcanism, but did not elaborate on it.[13]

I came in with *Strange World of the Moon*[5] in 1959. In the main the book followed in the footsteps of my volcanic, or properly plutonic, predecessors, and attempted a partial rehabilitation of Pickering. But I was forcibly struck by the manifold consequences of low gravity, to which they did not appear to have paid sufficient attention. The low bulk density of the rocks would make water a relatively heavy liquid, which would generally tend to stay underground, all the more so as the recent radio investigation had shown the lunar 'subsoil' (if this term is applicable to the Moon) to be permanently frozen, thus constituting a *permafrost seal*, which would not generally be broken by an ascending water column, except in combination with magmatic or tectonic action. Thus the existence of large underground bodies of ice and liquid water, perhaps even real seas, could be expected. These ideas were further developed in *Surface of the Moon* (1961).[6]

Geologically speaking, water and ice were incompetent rocks, and so could be compared to liquid rock, or magma. Thus sills, dykes and laccoliths of ice, injected either as such or as liquid water, could be important features of lunar 'geology'. Moreover, in the absence or great tenuity of atmospheric gases, ice would sublime without melting, and any liquid water would be technically superheated. Such superheated water would flash instantly into vapour upon reaching the sub-vacuum of the lunar surface and produce a *phreatic eruption*. Phreatic eruptions are not unknown on the Earth, as exemplified by the Tarawera blow-up in New Zealand in 1886, in which several craters up to 5 km[2] in area were blasted by superheated water. On the Earth, however, this requires high temperatures and pressures, and phreatic explosions are associated mainly with late-stage *ultravulcanian* activity when water comes in contact with magma. This can, of course, happen on the Moon as well, but the essential difference is that in lunar surface conditions liquid water will go off like gelignite even near freezing point. Further, an evaporating ice laccolith, known by the Eskimo name of *pingo*, would cause the overlying strata to cave in and produce a caldera. The effects would not be readily distinguishable from those of magmatic action, though neither need they exclude it.

Similar ideas were later put forward (whether independently or otherwise, is a moot point) by Z. Kopal, T. Gold[10] and H. C. Urey.[22] Indeed, this has resulted in something like an obsession with ice – almost a revival of H. Hörbiger's *Welteislehre*, originated by S. Ericson, of Norway, in 1885, subsequently developed by S. E. Peal, a British tea planter from Ceylon, and eventually accepted by P. Fauth in Germany, where it became an 'official Aryan doctrine' under the Nazis. Hörbiger's various extravagances included a Moon totally encased in ice, in which craters were formed by a process similar to Fountain's, mentioned on p. 105. In the sophisticated modern version as advocated by Gold (1966)[10] (for which, I dread to think, I may be partly responsible!)[5] only the maria are supposed to be underlain by a layer of ice about 1 km thick.

This has moved J. A. O'Keefe,[17] of Goddard Space Flight Center, U.S.A., to demonstrate that owing to the low rheidity of ice (p. 16) no craters over 1 km in diameter could maintain themselves in the maria for more than a few months if such a layer existed. Rilles would be similarly short-lived. His arguments, however, do not dispose of local ice pockets, as he concedes, pingos and other forms of water action, although he seems to think so.

We shall return to this problem in another context. Suffice it to note here that the main reason why some meteorists seized with such alacrity upon the idea of ice behaving like magma lay in the desire to escape the necessity of admitting vulcanism on the Moon, which is a tall order in the light of the *Orbiter* photography.

It all began a long time ago.

Chladni had just shown that meteorites were not a 'superstition of unlettered peasantry', after all : stones did actually fall out of the blue. And so it occurred in 1802 to M. von Bieberstein[15] that they must fall on the Moon as well, of which its craters were the visible sign and portent. Another German astronomer, Franz von Paula Gruithuisen, whose wild fancies put Pickering's 'small animals' completely in the shade, returned to this view in 1824. In 1873 it was popularized by an English astronomical writer R. A. Proctor,[20] though he appears to have recanted a few years later, his reason for this being that lunar craters were overwhelmingly round, whereas most meteorites would strike the Moon's surface obliquely and leave elongated scars, such as, say, the Alpine Valley (p. 71), claimed to have been sliced out by a tangential impact.

Nevertheless, G. K. Gilbert,[8] in the U.S.A., returned to the meteoritic (or meteoric) hypothesis in 1893 and gave it scientific status chiefly on the plea that lunar rings were too large to be volcanic. The main impetus to its further development, however, came from the recognition that the ¾-mile Coon Butte Crater in Canyon Diablo, Arizona, had been excavated by a large meteorite, now dated at about 50,000 years back. Curiously enough, however, when it was examined by Gilbert himself he pronounced it volcanic![15]

Yet the difficulty remained; the shape of the impact scars apart, the meteorite was supposed to dig the hole mechanically, like a bullet hitting a brick wall, and lunar craters seemed to be far too large for that. On the other hand, the meteorite (or was it a comet?) that hit Tunguzka, in Siberia, in 1908, produced a terrific blast, but no crater.

It was left to A. C. Gifford[7] to put two and two together in 1924 and recognize that the main excavating force in meteoritic impact was the explosive decompression of hot gases at the head of motion. Thus a meteorite was not so much a bullet as a high-explosive shell, and even an oblique hit would generally yield a round crater. In a later elaboration of the idea (1955) T. Gold[9] calculated that when the momentum of a meteorite moving at 50 km/sec was suddenly converted into heat at impact the resulting temperature would be of the order of $10^{7\,\circ}C$, as in the interior of the Sun or the fireball of a nuclear explosion (shades of Ocampo). Thus most of the meteorite itself and of the ground at contact would be instantly vaporized. The assumed velocity was admittedly a little high, but it was clear that the crater would exceed many times the diameter of the impacting mass, especially under low gravity and in relatively fragile lunar rocks. And true enough, for instance, the Sedan nuclear explosion pit in Nevada, some 400 metres in diameter, displays in miniature some of the features of minor lunar craters.

In fact, the inspiration for the modern version of the meteoritic hypothesis came not so much from astronomy as from the war-time experience of bomb craters. R. B. Baldwin, a writer and thinker of undoubted talent, based his reasoning on a graph where depth was plotted against diameter, from bomb craters through the few meteoritic craters known on Earth up to the medium-sized lunar rings, with statistically acceptable scatter in a single continuous line. The agreement broke down for the larger walled plains, but this was explained by their great age, erosion, isostatic adjustment, etc. He argued the case most persuasively in three books: *The Face of the Moon* (1949),[1] *The Measure of the Moon* (1963)[2] and simply *The Moon* (1965),[3] and gained many influential followers, especially in America, but he was not so much of a prophet outside his country, where his optimistic concluding remark that lunar craters 'can only be of meteoritic-impact origin. The 136-year argument is over',[3] was generally taken with a pinch of salt.

So it came about that on 21st August 1964 Cornell Aeronautical Laboratory, Inc. issued a very 'official' report *No. VS–1985–C–1*[19] on 'Some morphometric properties of the lunar surface', prepared by R. J. Pike, Asst. Operations Analyst, reviewed by W. F. Wood, Operations Res. Dept., approved by R. M.

Stevens, Head Operations Res. Dept. (how could you ask for more?), in which not only every crater, but even Oceanus Procellarum, deemed 'circular' and assigned a diameter of 1,500 km, is described axiomatically as 'an impact feature'. 'Orders is orders' and exactly the same attitude was truthfully portrayed in the statements of the three astronauts, Col. Frank Borman, Major William Anders and Capt. James Lovell, during the Christmas-1968 circumlunar flight of *Apollo 8*. Theirs was, of course, a stupendous achievement, nor can it be denied that the *C.A.L. Report No. VS–1985–C–1* contains useful data; but, while the astronauts, being army officers, may be excused, the authors of the *Report* are supposed to be pledged to scientific objectivity and cannot be let off so lightly.

A point that is often overlooked in planetary science is the necessity of controls, arising from the fact that any organic whole is more than the sum of its parts, and a complex physico-chemical system in which balance is achieved between various contending forces and processes over a long period of time is an 'organic whole' within the meaning of this definition.

If a biologist intends to test, say, the effect of red light on plant growth, he will take several, as nearly as possible identical seedlings of one or more species and grow some of them in red light and others in white light of the same intensity as *controls*. Without the latter his experiment would be meaningless.

At first sight this procedure may appear inapplicable to astronomy: we cannot make experiments with planets where one or more factors are varied at will, to see what happens. If we take it literally this cannot be denied. Nevertheless, small scale experiments of this nature are possible. The only difficulty lies in their validity when trying to predicate about so large an entity as a planet. Certain factors such as Martian or lunar gravity, cannot be duplicated in a terrestrial laboratory; others are not even known with adequate accuracy. Properties of the materials used in geological experiments may have to be scaled up in inverse proportion to the reduction in size and time,[12] which is very difficult and sometimes impossible.

However, the Earth provides a natural control, a touchstone of fitness, for theories spun about other similar bodies, and some 'mental experimentation' is possible. In this a self-consistent 'organic whole' is created in imagination from the observational data by a painstaking application of the known principles of science. This is a rather delicate art, involving such imponderable faculties as an ability to connect seemingly unrelated elements (Koestler's 'bisociation'), to visualize the implications of simple numerical data, and lastly, a measure of prophetic insight. By varying the assumptions made on the way we can then try to bring our conception into full coincidence with what is known, or assumed, to be the true observable situation. Sometimes this can be done mathematically, by manipulating one or more parametres within permissible limits, and we then speak of a *model* or a family of models. The difficulty of using the mathematical procedure successfully arises from the fact that it is at root analytical and so destroys the 'wholeness', or *gestalt*, to use a psychological term. With a large number of interlocking relationships the equations become unwieldly, and, there being more unknowns than equations, the result is unreliable.

For these reasons I am always suspicious of the usual technique of making a few selective assumptions, thought plausible, but plucked arbitrarily out of a complicated context, and leaving the rest to the march of mathematical reasoning, which takes care mechanically of the intermediate logical steps. This may produce a pretty academic paper, but its 'rigour' is often spurious and may be highly misleading. And in any event the conclusions are implicit in the assumptions.

Baldwin's simple empirical relationship between crater depth and diameter is not quite in the latter category, but his use of it is very much a case in point so far as the lack of controls is concerned. He never bothered to include volcanic calderas in his investigation. When this was done by P. Moore (1956)[14] and J. Green and A. Poldervaart (1960)[11] it transpired, however, that volcanic calderas fitted his graph equally well as the minor lunar craters and indeed better than the few terrestrial craters of established meteoritic origin.

On such evidence the craters of the Moon would be equally likely to be impact *astroblemes* and collapse calderas, nor could there be any compelling reason why both types of structure should not be present side by side, and quite obviously the graph can teach us nothing about other possibilities which have not been taken into consideration. To decide the issue, we must have recourse to other criteria, which will be reviewed in the next chapter.

REFERENCES

1 BALDWIN, R. B. (1949). *The Face of the Moon*. University of Chicago Press.

2 BALDWIN, R. B. (1963). *The Measure of the Moon*. University of Chicago Press.

3 BALDWIN, R. B. (1965). *The Moon*. McGraw Hill, New York.

4 BÜLOW, K. VON, GREEN, J., and ARMSTRONG, G. C. (1963). 'A Geometric Terminology of the Structural Features of the Moon', *Eclogae geologicae Helvetiae,* **56** (2), pp. 853–859.

5 FIRSOFF, V. A. (1959). *Strange World of the Moon*. Hutchinson, London.

6 FIRSOFF, V. A. (1961). *Surface of the Moon*. Hutchinson, London.

7 GIFFORD, A. C. (1925). 'The Origin of Lunar Craters', *J.B.A.A.,* **36**, p. 84 ff.

8 GILBERT, G. K. (1893). 'The Moon's Face', *Bulletin of the Philosophical Society of Washington,* **12**, pp. 241–292.

9 GOLD, THOMAS (1955). 'The Lunar Surface', *M.N.R.A.S.,* **115** (6).

10 GOLD, THOMAS (1966). 'The Moon's Surface' in *The Nature of the Lunar Surface* – Proceedings of the 1965 IAU–NASA Symposium, edited by W. H. Hess *et al.* The Johns Hopkins Press, Baltimore.

11 GREEN, J., and POLDERVAART, A. (1960). 'Lunar Defluidization and Its Implications', *International Geological Congress,* XXI Session, Part 21, pp. 15–33.

12 HOLMES, ARTHUR (1965). *Principles of Physical Geology,* Revised Edn., p. 240 ff. Nelson, Edinburgh and London.

13 JEFFREYS, SIR HAROLD (1952). *The Earth,* 3rd Edn. Cambridge University Press.

14 MOORE, PATRICK (1956). *The Origin of Lunar Formations*. Lecture to the Manchester University Astronomical Society.

15 MOORE, P., and CATTERMOLE, P. (1967). *The Craters of the Moon*. Lutterworth Press, London.

16 NASMYTH, J., and CARPENTER, J. (1874). *The Moon*. London.

17 O'KEEFE, J. A. (1968). *Water on the Moon and a New Non-Dimensional Number*. Preprint.

18 PICKERING, W. H. (1903). *The Moon*. Doubleday, New York.

19 PIKE, R. J. (1964). 'Some Morphometric Properties of the Lunar Surface', *Cornell Aeronautical Laboratory Report No. VS–1985–C–1.*

20 PROCTOR, R. A. (1873). *The Moon*. London.

21 SPURR, J. E. (1944). *Geology Applied to Selenology – The Imbrian Region of the Moon.* Science Press.

22 UREY, H. C. (1967). 'Water on the Moon', *Nature,* **216**, pp. 1,094-1,095.

Abbreviations used

J.B.A.A.   = Journal of the British Astronomical Association.
M.N.R.A.S.=Monthly Notices of the Royal Astronomical Society.

10

Astroblemes, subsidences and explosions

The approach to the problems presented by the mountain rings and craters of the Moon has often been somewhat naïve and lacking in scientific circumspection. To pick up the argument of the preceding chapter, it is clear that a sufficiently large and sufficiently fast meteorite can liberate at impact adequate energy for excavating a hollow comparable to the largest lunar walled plains. This, however, only means that the proposition deserves serious consideration.

There certainly exist erratic asteroids which could satisfy the energy requirements, but we cannot say *a priori* that they have ever hit the Moon, or hit it often enough, to produce the required effects. It must also be determined if, to what extent, or in what proportion the lunar craters have the character expected of impact features. We have already seen that the depth/diameter ratio is useless as a test.

The first necessity is, of course, to refer back to the Earth as the touchstone of fitness.

In 1924 D. M. Barringer carried out an intensive investigation of the Coon Butte Crater, since renamed after him, and established its meteoritic origin beyond reasonable doubt. Fragments of meteoritic iron were found in fair abundance; the crushed condition of the rocks indicated a violent impact. These findings made a great impression on scientific opinion, and a search began for other meteoritic craters, more particularly in the stable geological shields, where these would have escaped obliteration by later land movements, sedimentation or rapid denudation.

Canada yielded most of the candidates. These were (C .S. Beals, 1958)[2] the Chubb or New Quebec Crater, 3·4 km in diameter, classed as Recent; The Talemazane Crater, 1·8 km, late Pliocene; the Brent Crater, 3·7 km, early Paleozoic; and the Deep Bay Crater, 13·7 km, late Mezozoic. Baldwin reckoned the 120-km 'Vredefort structure', in Orange Free State, South Africa, and the 27-km Rieskessel in Germany among the astroblemes. The latter, however, shows clear signs of volcanic action, and competent geologists reject its meteoritic genesis.[21] This applies also to the Bosumtwi complex in Ghana and more particularly to the circular 52-km depression of Guelb er Richat in the Adrar Mountains of Mauretania, which the meteorists have also tried to claim in support of their theory. As for the Vredefort ring, this is according to W. H. Bucher (1965) only one of the five domed structures strung out in a straight line, 330 km long, and so most unlikely to be due to impact. Even the Chubb Crater fits the general tectonic pattern of the land (K. L. Currie, 1965), and is at best a doubtful item.[28]

The criteria themselves employed to establish meteoritic origin are not wholly unambiguous. These are the presence of cemented 'rock flour' known as *lechatelierite*, the impact form of silica called *coesite*, and of the stress structure of the rocks described as *shatter cones*. Yet volcanic explosions, too, can be of great violence (Krakatoa, Katmai)[39, 41] and may produce similar effects. At any rate, W. E. Elston (1965) found in the volcanic tuffs of Cerro Colorado something that looks suspiciously like shatter cones.[7]

The undoubtedly meteoritic craters are barely comparable in size to the volcanic vents, and even the Deep Bay Crater, if meteoritic, is easily outmatched by numerous volcanic collapse calderas. But we must also consider ring-dykes as the eroded relics of ancient cauldron subsidences. In 1963 A. R. Crawford discovered such ring-dykes up to 90 km in diameter certainly and 170 km probably in a pre-Cambrian complex of South Australia.[3] In 1965 W. E. Elston[6] described a volcanic structure, involving an encircled subsidence and bearing a strong morphological resemblance to the lunar ring mountains. It is represented by the 145-km Mogollon Plateau and the smaller Valles Caldera, 25·5 by 29 km, both in New Mexico. If eroded to the base, these would look very much like Crawford's Australian ring-dykes.

Thus Gilbert's argument from size falls to the ground. Even on the most optimistic reckoning, in the light of reference to the Earth the meteoritic theory presents no advantages over the volcanic-tectonic alternative. This does not eliminate the possibility that all or some of the lunar rings are astroblemes, but it makes this less likely.

The meteoritic conception has been closely linked from the start with Chamberlin and Moulton's *planetesimal hypothesis* of the genesis of the Solar System. According to the latter the planeto-genic cloud first condensed into numerous bodies of meteoritic and asteroidal dimensions, called *planetesimals,* which then aggregated into larger masses, and these picked up the remaining planetesimals. If, therefore, the Moon were an inert body, devoid of internal forces, and substantially immune to erosion, it was argued that it should have preserved more or less intact the marks of planetesimal impacts from the last stages of development of the Solar System.

This is not borne out by the evidence of telescopic observation and especially of the *Orbiter* photographs. We have seen in Chapter 8 that vast land movements have taken place on the Moon. The close-ups show lava and ash flows, leaving no doubt that powerful volcanic and tectonic forces have been at work. The craters are not all of the same age, and, slow as lunar erosion may be, its story is plainly writ on the Moon's face.

The planetesimal hypothesis itself, while plausible at first sight, encounters the basic difficulty that gas and dust do not preferentially coalesce into small, but into large masses. The present-day asteroids and meteoroids have not originated in the suggested way. If they had their density would have been very low, owing to the absence of compacting gravitational strain. Contrariwise, their structure bears witness to great heat and pressure of formation. They are fragments of a large body or bodies. As the French geologist S. Meunier has put it, 'the meteorites present a complex of stratigraphical relationships and by correlating these one can reconstruct a geological whole which bears a strong resemblance to our globe'.[4] The asteroids are often grouped in so-called Hirayama families, whose orbital evolution can be traced back to a single starting point (Brouwer and van Woerkom).[35] They must thus be products of the break-up of larger masses. The existence of polymict meteoritic breccias, in which fragments of meteorites of different types occur cemented together, clearly shows that the process of fragmentation has progressed by stages. The most economical assumption is that there once were two planets between Mars and Jupiter, and that the powerful perturbations by the latter brought these planets into collision. Such a collision would produce a violent explosion through the relief of internal pressures and the resulting fragments would become widely scattered in space.

The radioactive ages of the meteorites, reckoned from the time of formation, may generally be comparable to that of the Earth ($4\frac{1}{2}$ aeons), but if these ages are calculated from the helium content attributable to exposure to cosmic rays during the existence of meteorites as independent bodies they come up to no more than 300 – 500 million years.[36] This is the approximate age of the Brent Crater (Plate 12a). (The partisan objections to this finding by the adherents of the impact hypothesis of lunar craters need not be taken too seriously.) It would be rash to maintain that there were no meteorites before then, but any conclusions as to what meteoritic bombardment the Moon might have sustained before the Paleozoic Era based on the present situation must be suspended in thin air.

One criterion of distinction between endogenic (volcanic or other) and exogenic (meteoritic) craters is that the latter must have random distribution, once the fact that some primary crater-lands have been wiped out by the maria is taken into account. The only further local inequality theory permits – and indeed demands – is that the side of the Moon leading in its orbital motion should receive more impacts than the following side (Fielder).[9]

It will be observed straightaway that the maria are not distributed at random. We have further noted in Chapter 7 that the 'farside' thalassoids are concentrated towards the poles, which is the opposite of what one would expect from the sweeping by the Moon of interplanetary debris. This is paralleled by Fielder's finding that on the Earthward side craters between 4·20 and 4·79 km deep are heavily concentrated towards the south of the zone 20° W and 20° E. He has also found that the following side of the Moon is more heavily cratered than the leading side. (This is not necessarily significant and may be the result of the Orientale disturbance [see pp. 161 ff].).[9]

Crater statistics indicate clustering, but how much significance need be attached to this in the present context is an open question. Meteorites, too, occur in clusters; 'processions' of fireballs have been recorded; in the lower range of diameters secondary impacts by ejecta may have to be considered. On the other hand, the method of counting the number of craters per unit area does not reveal the existence of the well-known crater chains; indeed, it cannot distinguish between regular patterns and chaotic clusters.[10, 33]

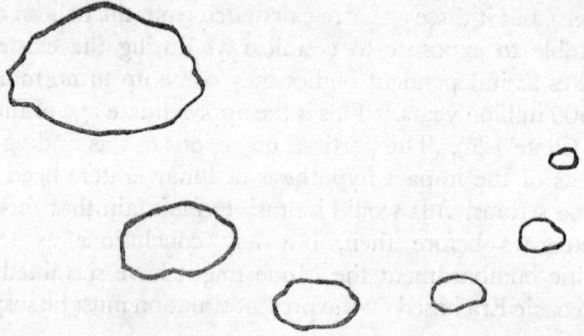

Fig. 5. Arcuate progression in Clavius (the largest crater
is Rutherfurd).

Far more important are the regularities in the distribution of craters of a given type, age and appearance, which readily catch the eye. When I first prepared rectified (spherical projection) photographs of the Fourth Quadrant I was forcibly struck by the fact that most medium-sized craters of clear-cut appearance stood in linear or arcuate arrays (cf. *Surface of the Moon*,[14] Plate I). Among progressions, where the comparable craters are not only aligned, but show a systematic decrease in size in one direction, mention may be made of the linear progression Fracastorius-Piccolomini-Stiborinus-Wöhler, the arcuate progression in Clavius, starting with Rutherfurd as the largest member, and the Halley-Hind chain north-east of Albategnius (6 and 5 craters respectively). (Figs. 5 and 6.) Further examples will be found in

Fielder's *Lunar Geology*[9] and Moore and Cattermole's *The Craters of the Moon*.[28]

Also significant is the concurrence of crater patterns with the grid system. This is very noticeable in many parts of the averted hemisphere. In the maria and the areas of low crater density similar craters often stand in triangular or rhombic groups,[14] which is readily understandable if they mark the intersections of the grid lines of tectonic weakness, as volcanoes often do on the Earth.[5,39]

Fig. 6(a). Halley-Hind arcuate progression. (b). Arcuate progression near Condamine (dashed lines).

The walls of large lunar rings are usually conspicuously aligned with the grids as well, of which Ptolemaeus and its neighbours provide a classical example. Another important point is the contiguity of such rings as Ptolemaeus and Alphonsus, which have one wall in common, without any overlap. Kopal, who is otherwise favourable to the impact hypothesis, considers it improbable that this effect could be due to chance, and regards such craters as endogenic.[24] This, though, is only a matter of probability, and improbable things do happen from time to time, an objection that is invalid in the cases where contiguity involves deformation of outline, as exemplified by the 'beetle' patterns of Heraclitus and Hainzel. We have here a cellular arrangement, as though the growth of the constituent rings has been inhibited by their neighbours.* A striking example of such contiguous structure will be seen in the frame HR 034, 3 of 3 of the *Orbiter II*. Cellular patterns of squarish rings also form the backbone of the Leibnitz Mountains and some heights south-east of Mare Australe.

* Similar patterns occur among the explosive maar craters in the Queen Elizabeth National Park, in Uganda (Plate 12b).

Such patterns cannot be produced by random impact, and some formations involved in them exceed 50 km in diameter (Fig. 7).

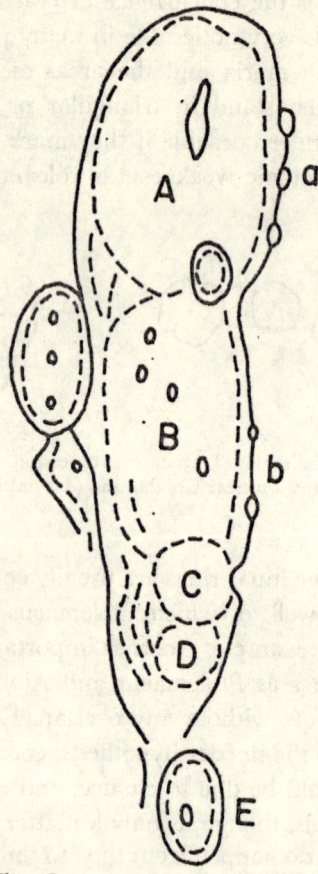

Fig. 7. The five components of Palitzsch. E is probably a later structure, but clearly lies on the same structural line. The small craters at a and b have formed on the crest of the rampart. (Reproduced from *The Craters of the Moon* by Moore and Cattermole with the authors' permission.

Ditch-like confluent craters, such as Schiller, Vögel or Palitzsch, present a similar problem. Genetically very significant is also the case of what Moore[28] has called (1966) 'interlocking craters', where a crater bestrides the wall of another, usually larger, ring and this wall is still traceable across the floor of the overlapping crater. The already-mentioned Rutherfurd on the wall of Clavius is a good example of this. The detonation accompanying a meteoritic impact would certainly have obliterated any trace of a pre-existent structure. Indeed, it must be expected to produce a violent shock wave that would shake and shatter the surface features over a considerable radius around the point of impact. Yet, as has been noted by both Moore and Kopal,[24] the crests of the crater walls at the points of intersection of overlapping craters usually stand clear and sharp, as though cut off with a knife, which is again incompatible with explosive origin, whether volcanic or meteoritic.

Crater overlaps have not been statistically reviewed, but there is a general rule that the overlapping crater is smaller than the overlapped one. A classical illustration of this is provided by the crater Thebit in Mare Nubium, where three craters overlap coaxially in diminishing progression. Examples of this relationship are legion. There are exceptions, but they are so few that they may be said to 'confirm the rule'. T. Sato[34] and associates, of Hiroshima, Japan, made a thorough search for these on the Earthward hemisphere (1968), but have been able to find only some 30 'forbidden overlaps' among hundreds of 'normal' ones, and some cases cited by them may be a little doubtful.

A clear instance of a larger crater encroaching upon a smaller one will be seen about halfway between Plato and Mt. Blanc off the north 'shore' of Mare Imbrium. The *Orbiter III* photographs (see Plate 16) show two conspicuous 'forbidden overlaps' north of Tsiolkovsky on the 'farside'. One of these is of further interest inasmuch as the larger crater has invaded the smaller one with an unmistakable flow of lava, so that the crater must be taken to be of volcanic origin.

The great rarity of 'forbidden overlaps' is, nevertheless, significant; for, while a few such must be expected on any assumptions, it is clear that a declining internal force acting in the same area will tend to result in craters of progressively decreasing size. If, though, the craters are predominantly exogenic such a sequence cannot be expected to be strictly observed. It may be true that smaller meteorites are more numerous than larger ones, and that the average size of meteorites has declined with time; but meteorites and asteroids of all sizes exist now, and there is no reason to assume a uniform decrease in the force of successive impacts.

Craterlets below 1 km in diameter may have to be considered under a different heading, but their distribution is extremely uneven. There are generally more of them on level ground than on the hills, and more on or near the ridges and summits than on the open slopes. The relative age of lunar formations is not always easy to assess, but in examining (1965)[16] the *Ranger IX* photographs of the Alphonsus region of the Ptolemaean highlands I have found that there are about ten times as many craterlets on the floor of Alphonsus as in the neighbouring part of Mare Cognitum. Ptolemaeus seems to be older than Alphonsus and Albategnius than Ptolemaeus, but the density of craterlets on their floors is in the inverse order of age, the respective ratios being approximately Alphonsus/Ptolemaeus/Albategnius as $8·5 : 6 : 4$. Moreover, the craterlets in Alphonsus and Ptolemaeus occur in preferentially oriented chains and their trends differ in the two walled plains.

The Alpine Valley (p. 71) is selenologically younger than the uplands it has parted. If, therefore, craterlet density were simply related to the number of meteoritic impacts it should be higher on the older surface than on the younger: precisely the opposite is true (Firsoff, 1968).[18]

Small craters and craterlets often wreath the walls of large rings, of which Gauricus is a good example. This does not fit the impact explanation either.

Often invoked in its support is the approximate rule formulated by Schroeter, whereby if the rampart and outer glacis were bulldozed down to a dead level the material thus gained would exactly fill the hollow within. This may seem plausible at first sight. Yet the volatilization of the impacted ground at temperatures that may reach $10^{7°}$ C,[19] the ejection of debris to great distances and its consequent loss to the balance sheet, and, above all, the compacting effect of the violent shock on the light and porous lunar foamstones, are completely overlooked. If these are taken into consideration it will be seen at a glance that the volume of the excavation must be much greater than that of the ejecta around it.

In fact, if Schroeter's rule were strictly observed it could with better reason be cited against impact and in favour of volcanic origin.[41] For, assuming that the material of the walls and outer glacis were derived from a magmatic chamber beneath the crater floor, which then subsided into the vacated space, the volumes of the two should balance.[41] Nor would the objections just mentioned apply, or apply to the same extent (Fig. 8).

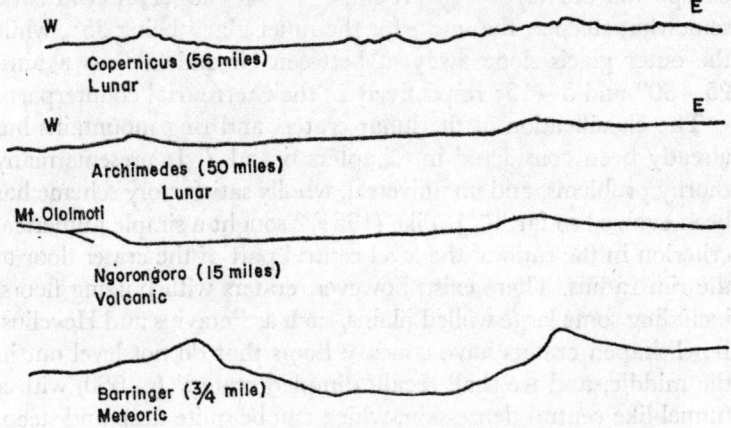

Fig. 8 Sectional profiles of lunar and terrestrial craters.

Alas, the rule is not even approximately correct. Fielder put it to the test on six craters, using the data compiled by the members of the Lunar Section of the British Astronomical Association, and found that in every case the volume of material outside the depression exceeded its volume from 2 to 7 times. The factor was 7 for the craters, Euclides, Cayley, Hortensius and Nicollet, with diameters from 11 to 16 miles. Measurements of 39 profiles of larger rings on the U.S. Air Force charts have again shown excess of the rim volume over that of the hollow.[9]

Spurr (1944)[37] was the first to note that such craters as Autolycus and Aristillus sat to top of gentle upswellings, or shield structures, and B. Warner found that in the case of craters of the order of 10 km this shield structure was between 3 and 4 times larger in diameter than the crater.[9]

All this militates against the meteoritic hypothesis.

Furthermore, craters of the same diameter differ greatly in depth, and Baldwin could not have established his celebrated relationship without recourse to selection, often based on arbitrary and suspect interpretation (Pike, 1964).[33] A more sophisticated plot, in which six types of lunar craters are considered separately, was prepared for about 320 objects by the Russian astronomer G. K. Katterfeld (1966)[23] and compared with 35 terrestrial calderas. The scatter of the depth/diameter ratios is very similar, except that the lunar rings are some 10 times larger; they are also somewhat steeper, the mean for the inner glacis being 35°, while the outer glacis slope away at between 10 and 20°, as against 25 – 30° and 5 – 15° respectively in their terrestrial counterparts.

The classification of the lunar craters and ring mountains has already been considered in Chapters 6 and 7. It presents many thorny problems, and no universal, wholly satisfactory scheme has been evolved so far. R. J. Pike (1964)[33] sought a simple numerical criterion in the ratio of the level central part of the crater floor to the rim radius. There exist, however, craters with bulging floors, including some large walled plains, such as Petavius and Hevelius. Bowl-shaped craters have concave floors that do not level out in the middle, and we shall recall 'dimpled craters' (p. 000) with a funnel-like central depression, which can be quite deep and steep. For instance, D. W. G. Arthur[1] further distinguishes ring-craters with small dark floors and small bright floors. Larger craters may also be classified according to the shape of the walls, whether single or multiple, terraced or grooved. The presence or absence of central peak or peaks, their size and height, provide other distinguishing marks, to say nothing of domes, craters, rilles, lava flows, etc., on the crater floor, rays and haloes outside the ring (Plates 4a, b and c). Changes of appearance due to age should not be forgotten either, and, though clear in extreme cases, may be open to various interpretations: for instance, the absence of a central peak is often held to be a sign of age, yet some old, eroded rings preserve their central peaks, while clear-cut rings of youthful appearance have none.

Such morphological differences may or may not be of genetic significance. There exists a correlation between the size and structure of craters, with a considerable overlap at the top and bottom of the scale. Dimpled and bowl-shaped craters seldom exceed 10 km in diameter, and it is statistically accurate to say that terraced walls and central peaks are the prerogative of larger rings. There is a special type of bowl-crater, so neat and accurate 'as though it were turned out on a lathe'. The shape seems, further, to be gravitationally determined, as the rims conform to a horizontal plane, regardless of slope. Such craters are particularly common in Oceanus Procellarum and around Copernicus. There are ragged and irregular craters as well.

Clusters of small craters which resemble knuckle marks made by a fist on hardening plasticine, for instance, in Sinus Aestuum and radial to Copernicus, are most probably secondaries, produced by ejecta due to primary meteoritic impact or volcanic bombs.

In the latter connection we may usefully refer to the *Surveyor I* frame 66–1910 (Plate 19b), showing at about $3\frac{1}{2}$ metres an elongated lunar boulder 45 cm long. The appearance of the boulder recalls a piece of dough, which was flung along at no great speed when already hardened on the outside, but still plastic within, the inner mass having surged forward after the fall, causing the skin to wrinkle. This describes a volcanic bomb, as has been decided quite independently by G. P. Kuiper[25] and myself (1967).[17] Other rocks in the close-ups taken by this probe cannot be examined in equal detail, but their pear-drop shape and speedboat attitude suggest the same nature.

This may be taken to mean that some explosive vulcanism has occurred on the Moon at an unknown date. Seeing, however, that the terrain is coated with a seemingly unconsolidated material (see p. 236), which may be volcanic ash, to a depth of a few millimetres, and this material appears to be absent from the boulder and the fragments broken off from it, the eruption in question could not have been very ancient.

At this stage we may again recall those dimpled craterlets photographed by the *Rangers, VII, VIII* and *IX* in the Maria, Cognitum and Tranquillitatis and on the floor of Alphonsus, but also present in the region of Tsiolkovsky and around Mare Orientale, where they may attain dimensions of up to several kilometres. Kuiper was greatly impressed by their resemblance to gas-escape cavities in some lavas seen from the air (1965), but the latter look harsher in outline and less regular than the soft-shouldered lunar depressions, and I tend to side with O'Keefe[30] in thinking that the lunar material is ash, not lava. However, such cavities could also be drainage holes, of *swallets* (Firsoff, 1959;[13] Kuiper, 1965[17]). One thing they cannot be, and this is impact features.

Neither explanation is applicable to those dimpled craterlets and craters which are perched on top of ridges and conical peaks. About 200 km south of Tsiolkovsky there is a broken lava-filled enclosure, some 100 km wide and sporting a central dome. Hard below the south wall of this ring, at approximately 36° S and 125° E, rises a shapely cone terminating in a dimpled crater, about ½ km in diameter. Clearly, this is a conventional ash volcano. Such structures are by no means uncommon on the Moon.

While in this area, we may examine frame 121 (2 of 3), taken by the *Orbiter III* and showing a rim of Tsiolkovsky and a 60-km ring immediately to the south of it. A geological eye cannot miss the marks of magmatic activity in Tsiolkovsky itself, but the smaller ring is self-evidently part-filled with lava (and/or mud), which has partly come down from the sides and has partly welled up from below, raising botryoidal domes. A succession of flows, some of which, moreover, appear to have been partly washed away, are readily distinguishable. The character of the flows has changed over the period involved, which must have been considerable, some flows being richly peppered with shallow dish-like depressions, while others are more or less smooth, and, since the latter underly the former, they would appear to be older as well. Once more the density of craterlets is in inverse order of age, accordingly. There are, however, a few bright craterlets, which are probably meteoritic (Plate 9).

The flows bear a strong resemblance to Icelandic lava fields of highly fluid basic lava, but with a difference. The fronts of the Icelandic flows have fingered or dendritic outline, whereas here they are much more massive, some of the lateral effusions ending up in moraine-like mounds. This may be taken as further evidence of intense vesiculation on the exposed surface of the lava (p. 96).

Let us now move to the front side of the Moon and consider the *Orbiter V* views of Tycho. This rayed crater is often held forth as a model of lunar impact feature. Both, however, the floor and the outer glacis of the crater are caked with lava. The crater is also surrounded by an external collar, which recalls that of the terrestrial collapse caldera of Ololmoti in Tanzania[14] (Fig. 9). The Tycho lavas are very different from those just described; they are blocky, criss-crossed with crevices and scattered with tholoidal masses. Here, too, the development appears to show several phases, extending over a considerable period.[38] (A *tholoid* is a mass of solidified lava extruded from a vent or fissure by a rising column of magma. It is characteristic of high-viscous acidic vulcanism of the so-called Pelean type – after Montagne Pelée in Martinique. In lunar conditions, however, it is not always easy to decide whether the extrusion was in solid, semi-solid, plastic or liquid phase, and for this reason I have used this term in a generalized meaning for any extrusive structure.)

Such differences of appearance must reflect a difference in composition, temperature and perhaps atmospheric conditions. Inasmuch as Tycho is a bright lunaritic crater, its lavas may be expected to be acid.

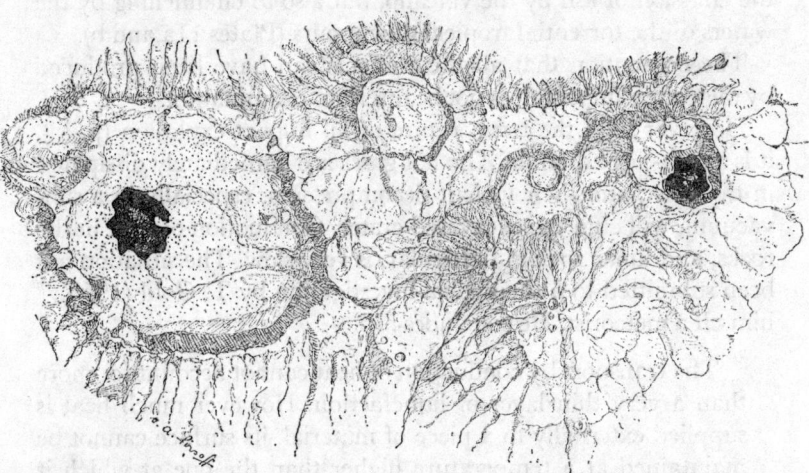

Fig. 9. Relief map of the Crater Highlands of Tanganyika. *Left to right:* Ngorongoro with the soda lake Magad, Ololmoti (see plate X), Mt. Olossirwa and Mt. Loolmalassin, Craters Embulbul and Elanairobi with Lake Embagal. Ololmoti is surrounded by a collar-like terrace which bears a close resemblance to the collar round the lunar ring-mountain of Tycho. (Drawn by the author from the maps and charts supplied by the courtesy of the Chief Surveyor of Tanganyika).

The oblique views of Copernicus and Theophilus, two comparable craters, obtained by the *Orbiters, II* and *III,* show soft landscaping, suggestive of ash. Their ray systems are also different from the bright streaks emanating from Tycho, and especially those of Theophilus are relatively dark and faint (p. 78). There is no indication of anything like the boulder-scree photographed by the *Surveyor VII* on the outer glacis of Tycho. Some differences of 'habit' are, therefore, not unexpected. Yet when the floor of Copernicus is viewed vertically from above in the *Orbiter V* frame

MR 156, its structure, though somewhat less rough, does not look very different from that of Tycho. On the other hand, there can be no mistake about the outer glacis of Aristarchus being shaped by ash flows, which feather out towards Schroeter's Valley in the *L.O. V* frame MR 202. Frame MR 196 is even more striking. Here the slopes of Aristarchus look suggestively like those of the Tanzanian volcano Oldonyo Lengai, seen from the air; and it will be appreciated that the latter owe their appearance not merely to the emission of ash by the volcano, but also to channelling by the waters of the torrential tropical downpours (Plates 11a and b).

The contention that all these effects can have been produced by meteoritic impact does not bear serious examination.

However high the temperature of an impact explosion (p. iii), it is a split-second affair. The hot gases are immediately dissipated in the blast, and their violent decompression, especially in a sub-vacuum, will chill the excavated crater. Conduction of heat in the rocks, and lunar rock in particular, is very slow. The situation has been submitted to a theoretical investigation by T. Gold (1955),[19] himself a meteorist. He concludes:

> The transport by conduction of heat cannot account for more than a very thin layer of liquefaction. However much heat is supplied externally to a piece of material, its surface cannot be maintained at a temperature higher than the one at which it vaporizes.

At most some superficial glazing may result: in no case lava, let alone repeated flows of lava or effusions of ash. This is amply confirmed by the actual experience with terrestrial meteoritic and nuclear explosion craters.

The impact mechanism is equally incapable of explaining such common features of lunar mountain rings as level floors, central peaks, multiple terraced walls and parasitic craters and crater chains.

If we have a look at, say, the *L.O. III* frame 78, the level-topped terraces at the nearer north-east rim of Theophilus are obvious dropped benches resulting from a central subsidence along concentric ring step-faults. In the south the terraces are inwardly bounded by linear ridges corresponding to fractures traceable beyond the ring.

On the other hand, the listed particulars are all present in laccolithic or other calderas, described by Howel Williams (1941).[41] The mechanism of formation may be summed up as follows: Magma (or other incompetent material)[5, 14, 20] is intruded from below and spreads mushroom-wise (laccolith) between the strata, uplifting the overlying rocks into a dome. If the uplift is sufficiently sharp and the rock brittle (as on the Moon, *see* Chapter 8) inwardly dipping concentric fractures will form. When invaded by magma, they are known as *cone sheets*. The subsequent withdrawal of magmatic support causes the dome to cave in, yielding a ringed depression. The withdrawal of support entailing a downward movement of the surface strata will also cause concentric fracturing, but of opposite inclination; such outwardly dipping fractures when injected with magma become *ring dykes* (Anderson, Billings).

The process is somewhat complicated and allows of several variants.

Among the causes of withdrawal of support Williams[41] lists (1) 'Loss of gas and contraction of magma on cooling . . ., (2) rapid outflow of lava from fissures far below the summit vent . . ., (3) rapid eruption of great volumes of new magma as pumice and ash . . ., (4) changes in the shape and volume of magma bodies at depth . . .' If the intruding material is a salt diapir, it may be removed by solution; if ice – by melting or evaporation[14] . . . And all this may happen *at depth*.

An important factor is the depth at which the laccolith is emplaced or the movement of magma takes place. In the case of shallow-seated intrusions, the magma may either reach the surface and produce a volcano, or it may stay sealed and yield a closed dome or a blind *puy*.

Most of the terrestrial calderas are volcanic and due either to (2) or (3). The floor of such a caldera will generally be elevated above the surrounding country. Some lunar craters, at the edges of mountain chains, such as the Rooks, the Caucasus and the Altai, and on the mountains at the far end of Mare Australe described in Chapter 7 (p. 86), conform to this pattern. In most cases, however, their floor is depressed below the country level. Yet there exists a class of terrestrial vulcanism of type (3), where a single eruption blows out a shallow, low-walled depression with a substantially flat bottom known as a *maar*. In the course of time most maars come to cradle a lake, as the level of their floors usually lies below that of the surrounding country. Their appearance closely parallels that of minor lunar dish craters.

Maars are well known from the Eifel district of Germany, south-western Uganda (Plate 12b) and Kamchatka. The already-quoted Katterfeld's article in *Icarus*[23] contains instructive reproductions of some aerial photographs of Kamchatkan maar craters.

If the dome remains sealed, its caving-in is usually due to (1), and results in a *crypto-volcanic depression*.[5, 21] Such structures as Rieskessel, Vredefort, Guelb er Richat, seem to fall in this category. Once more the floor of the depression will typically lie below the surroundings.

An energetic upthrust close to the surface tends to induce radial fracturing, characteristic of many domes, and the radial gullies of such lunar craters as Aristillus, Autolycus or Timocharis,[37] which are seated on top of gentle domes (p. 125), may be of a similar nature. Indeed domes are extremely common on the Moon, for instance, in the neighbourhood of Hortensius, and they often have a central crater. Within some large rings there are many more domes than small craters; while in the old formations of the same size domes are usually absent and the floor is richly cratered. It seems logical to conclude that there is a connection between the two, and the latter craters and craterlets are due to the collapse of the erstwhile domes. In the case of the overlapping craters mentioned on p. 123, the lava has descended from the larger and deeper into a smaller and shallower crater, and it is difficult to see why this should be so unless the former had not been there at the time.

Of special lunar interest, however, are *cauldron subsidences*[22] along the ring fractures, arising from cause (4), which may be mediated by the injection of dykes or sills, and in the case of the Moon perhaps tidal movements of magma due to the varying attraction of the Earth. Scottish cone sheets dip inwards at about 45° and when produced downwards meet at a focus $4 - 5$ km below the surface.[5] It will be appreciated, however, that the size of the ring defined at surface by such a cone sheet will depend on two factors: (a) The depth at which the apex of the cone lies, (b) The angle of the dip.

From the considerations of lunar tectonics (Chapter 8) and the expected honeycomb structure of the lunar crust fractures should generally penetrate much deeper on the Moon than on the Earth.

Further, the dip angle depends on the coherence and rigidity of the rocks, and so should be smaller in lunar conditions (*see* Fielder's *Lunar Geology*[9]). It is clear that in combination lunar cone sheets will describe at emergence much wider circles. From Katterfeld's[23] comparison between the depth-diameter ratios of the terrestrial and lunar calderas (p. 126) it seems that the average difference is by a factor of 10, which is not at all excessive. If now, bearing in mind Crawford's South Australian ring complexes (p. 117), we put the diameter of the largest terrestrial cauldron subsidences at 100 km (the Aleutian arc is a much larger structure), then there is no difficulty about interpreting even the most extensive marial basins in the same terms.[15]

We may also consider in this context the type of structure known as *lopolith* (from the Greek *lopas*, a shallow basin), a saucer-like depression of the ground which is pulled down isostatically by a deep-seated intrusion of heavy magma. The Bushveld lopolith in South Africa has a diameter of over 300 km (Holmes).[21] As adumbrated in Chapter 5, the maria may have been formed in a similar way.

In most cases ring dykes and cone sheets can be studied on the Earth only as ancient eroded structures, in which the land movements are clearly apparent, but the surface expression can be only indirectly inferred – somewhat like reconstructing the appearance of a vanished species from its fossil bones. In contrast to the shallower collapse calderas, the inner rock mass of cauldron subsidence tends to drop *en bloc,* without shattering, and so yield a level floor. Nevertheless, step-faulting along concentric fractures must be expected at the rim. The oppositely dipping annular fractures, produced in the upheaval and withdrawal phase respectively, will meet at a sharp angle, which should yield ridging at the edge of the terraces. This effect is well illustrated by Guelb er Richat, but also in the large lunar mountain rings, e.g. Theophilus. The collapsing rock block will be compressed towards the centre; and, since collapse induces annular fracturing with an outward dip, at the centre it will define a more or less steep cone, which will tend to be pushed upwards as a central peak. If we look at the *L.O. II* close-up of Copernicus, the mountain on the crater floor just left of the centre clearly displays vertical joints which diverge upwards, indicating that it has been formed by extrusion, as inferred above.

So far we have left any magmatic activity at the surface out of account, which is a possible variant where the magmatic phase is weak. In the case, however, of the Mogollon Plateau (p. 117), which represents a surface expression of cauldron subsidence, we have intense vulcanism both at the centre and along the bounding fracture, where lava and immense amounts of ash have been emitted, building up a mountain range.[6] This seems to be closely paralleled on the Moon.

In examining (1965) the *Ranger IX* close-ups of Alphonsus, and more particularly frame 3 : (65–1343),[16] I was forcibly struck by the fact that the east walls of the ring ran parallel to the floor fractures and enclosed between them what appeared to be an unaltered portion of the floor. In other words, the walls were posterior to the floor and have been built by fissure emission of some material, which seems to be volcanic ash. Other investigators (Kuiper, O'Keefe, Cameron, Fielder) have reached quite independently the same conclusion.

The independent investigation of the large fragmentary ring Flamsteed P in Oceanus Procellarum by Fielder (1967)[11] and by J. A. O'Keefe, P. D. Lowman and Winifred S. Cameron (1967),[31] based on the *Orbiter I* and *Surveyor I* photography, points in the same direction. The heights following the ring fracture are tholoids or effusions of highly-viscous lava, arrested in its progress by intense surface vesicultion (coacervaion, Fielder), as is clearly shown by flow-ridging and a bulge or welt, more learnedly described by Fielder as 'lobate terminus', at the boundary or foot of the hill (Plate 8b). This is a characteristic lunar form, found again and again – in the marginal hills of the Lunar Alps, in the group of heights near Herodotus, and elsewhere. Contrary to O'Keefe *et al.*, some of the Flamsteed heights are cratered; they also pass into wrinkle ridges. The latter are lunabase, but the tholoidal hills are lunarite.

The approach in the two papers differs considerably. Fielder appears to be greatly concerned with demolishing the idea that such rings as Flamsteed P or Lamont in Mare Tranquillitatis are ancient formations, submerged and destroyed in the marial upheaval (whatever this may have been), and holds forth contrariwise that they are young mountain rings in the process of emergence. He describes such formations as Egede, Wallace and Stadius as 'elementary rings' (p. 69), with a single wall equally steep on the in- and out-side and without a central depression, as a next stage in the process. There is indeed some evidence in support of this view. In Archimedes, however, we have a well-developed ring structure, which appears to have been invaded by marial material, and those rings which are fully formed on the terra side, but dip into the mare and are lost in it would point to destructive action.[14] This evidence cannot be dismissed, but neither is it vital to Fielder's conception, for rejuvenation of the old fractures in submerged rings may be expected and would still bear witness to the kind of marginal effusion he envisages.

The viewpoint of O'Keefe *et al.* is even narrower. They seem to miss the wider implications of their own findings, continue to regard mountain rings as impact features, and concentrate – to my mind excessively – on the conjectural composition of the lavas involved in the extrusion. Their use of the formula of Nichols and Jeffreys to demonstrate that this lava could not have been basic is a bad example of injudicious introduction of mathematics into planetary problems. They disregard vesiculation, put the density of basic lava at 3, which, when fed into the formula, gives a speed of 'steady flow' of 1 km/sec, in other words, 3,600 km/hour! It does not occur to them that such lava could be used as a rocket propellant . . . (*see* Appendix, p. 248).

The defenders of the impact hypothesis are forced into an impossible position, where the ring mountains arc basically impact features (on what evidence?), but modified by secondary vul-canism so as to look *like* collapse calderas or cauldron subsidences (they do not put it in this way, but this is what their argument amounts to).[29] It may be that in conditions of intense vulcanicity a violent impact may trigger off volcanic activity; but, being given these conditions to begin with, the impact is wholly unneces-sary to produce the observed effect.

From his findings in Flamsteed P and other 'elementary rings' Fielder has developed a theory of formation of lunar mountain rings that has much to recommend it.[9] It all begins with magmatic activity along a cone sheet, leading to the formation of an elemen-tary ring. As more magma rises to the surface, the fracture is further activated, with marginal emission of ash and lava, which results in a kind of ring volcano. This leads to the loss of material from the subjacent magma reservoir and a cauldron subsidence within the ring along annular faults of reversed dip. The sequence of events is illustrated in Fig. 10.

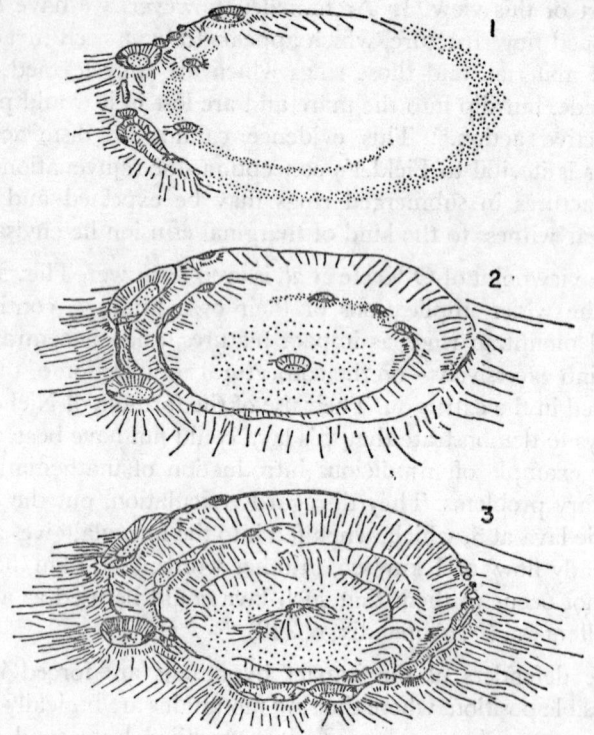

Fig. 10. Development of a lunar ring-structure after Gilbert Fielder: 1, early volcanic action in annuli defined by ring-fractures (shaded); 2, complete growth of an outer ringwall, and partial growth of an inner ring, or spiral, by volcanic extrusion. Some subsidence and melting of the floor of the main ring-structure; 3, further in-dropping along ring faults, further development of the outer and inner walls by lava flows from the rim volcanoes, and the replacement of a central orifice by central cones, lineamented in the direction of regional fractures. (From *Lunar Geology* by courtesy of Lutterworth Press)

As we have seen (p. 130), there is good evidence for, say, Aristarchus being a ring volcano, on the lines comparable to, if somewhat different from, those envisaged by Spurr.[37] Annular faulting and fracturing is clearly visible, e.g. in Inghirami (*L.O. IV* frame 172) and Plinius (*L.O. V*, HR 070–2 of 3). The Hadley Rille in the

Apennines starts from an unsealed ring fracture. The numerous examples of inner rings within craters, investigated by B. Warner (1961),[40] bear further witness to the soundness of the basic idea. Nevertheless, the difference between a cauldron subsidence and a collapse caldera is one of degree only, and the older conception that at least some lunar craters are of the latter type (p. 86) cannot be discarded. Others are best explained as explosive maars, and, though less conspicuous than the great rings, conventional volcanic cones are not absent either.

It remains to be asked, What about meteoritic craters? The Moon must possess its proportional share of these as well. Fielder thinks that every tenth crater may be meteoritic.[9] The proportion of impact pits among the small craters could be higher, but he seems to be over-generous if this is meant to refer to the large rings, for I have not found a single case which inevitably requires impact interpretation. Some small craters with bright haloes which close-up photographs reveal to be composed of bouldery material, ragged rims and bowl-shaped profiles must have been explosively excavated. The *L.O. V* MR 081 frame shows the 19-km Dionysius to be of similar structure and probably meteoritic origin. But an explosion does not necessarily imply impact.

The *L.O. III* photo No. 67–H–204 shows such a craterlet 150 metres (500 ft.) in diameter. It encloses a second concentric crater-let, the ratio of the rim diameters being 3·5 : 2·5. Nearby is a smaller, eroded or overlaid, depression of similar structure. The official caption reads, '. . . The double-walled appearance is due to a continuous landslide around the circumference of the crater . . . Blocks and boulders averaging about three feet in diameter which were ejected from the crater form a symmetrical pattern of rays . . .' Some of the ejecta conspicuously bridge the gap between the two walls, and so would have to have been ejected after the landslide! We may, however, refer to *L.O. II* frame HR 177 (2 of 3), which shows a very similar explosion crater, but within it there are several (not just one) smaller craterlets, side by side. Have these, too, been produced by some incredible landslide? This explanation is neither reasonable, nor indeed possible. It is quite clear that two successive concentric explosions have occurred in the first case, and three non-concentric ones in the second. These cannot have been due to impact; the most natural cause that suggests itself is phreatic eruption, resulting from an ordinary spring of water making its way to the surface, blowing up at contact with the sub-vacuum and then freezing (Plates 10a and b).[14, 18]

In any event shattered rock, including chunky boulders, is involved in at least some bright haloes, although it may not wholly explain their colouring. On the other hand, the light-tinted matter on and beyond the slopes of Aristarchus is substantially volcanic ash. The *L.O. V* frame MR 142 shows the same feathery lacing in the halo of Copernicus, which must, therefore, be of similar composition. In this case some ejection of lumpy material (volcanic bombs?) appears to have intervened as well, but it cannot account for such effects as a ray being stopped by a wrinkle ridge. Moreover, some ray elements (plumes, *see* p. 77) are clearly associated not only with ambiguous elongated depressions, which may be due to secondary impact, but with unmistakable surface fractures and chains of craterlets of internal origin (cf. *L.O. II* MR 144).

Fielder and Warner (1961) have studied the stress patterns,[12] expressed as fractures and ridges, around lunar craters, and found that these tend to be distributed in families, only the central line of which passes through the centre of the crater, the whole family being enclosed in an approximately elliptical 'envelope'. Now rays frequently follow the same pattern.

A local stress field around a crater is affected by the general grid system (p. 121), and there exist other stress lines of parallel habit, which though related to a crater cut right across it, becoming tangential on the periphery. This kind of fracturing is clearly visible in the *Orbiter* views of Tycho, where again the fracture trends coincide with those of the rays. This cannot be wholly fortuitous.

The colour photographs taken from the *Apollo VIII* give the halo material of many craters a peculiar chalky hue, which could hardly be due to the simple crushing of brownish or greenish rock. Some kind of deposit or incrustation seems to be involved in this. Pickering[32] thought it was snow, which does not appear to be feasible in the light of our present knowledge. I have suggested (1961)[14] ammonium chloride as a frequent volcanic deposit.

The length of the Tycho rays as such is no bar to their having been formed by ejection. Volcanic projectiles at emergence may move at speeds exceeding the velocity of escape from the Moon (2·38 km/sec), and the blast generated by a meteoritic impact is

equally up to the mark. It is, indeed, widely believed that the tectites have come from the Moon in the one or the other way.[26, 29]
The fact that several rays of Tycho, as well as those of Proclus
and Thales, are tangential to the rim of the crater is rather
awkward, however, even though tangential ejection seems to have
been observed in some nuclear explosion craters. Nevertheless, it
is to be doubted if simple ejection could result in a substantially
straight, more or less continuous streak extending to distances over
1,000 miles. There is also a curious association between the rays of
Tycho and other rayed craters. Thus a great double ray running
over Mare Nubium into Oceanus Procellarum is aligned with a
succession of bright craters, including Euclides, Encke, Kepler
and Aristarchus. Another ray seems to be 'picked up' by Menelaus
and Bessel in Mare Serenitatis.

This is only an outstanding example of a common occurrence,
where a ray is 'revived' upon encountering in its path a further
rayed crater as though it caught up and entrained something
emitted by the latter. Pickering suggested electrostatic repulsion
between ray particles as a possible mechanism. A steady jet of gas,
travelling like a kind of linear wind, which is conceivable in lunar
conditions, would do as well. But neither would work unless the
minor crater were in emission at the same time as the big one. In
other words, the concurrence of evidence points to activation of
tectonic lines of weakness by some event in the main rayed crater –
impact, eruption or some more gradual relief of stress.

The idea that the bright rays may be, or be associated with,
surface fractures appears to have been first proposed by Tomkins
in 1906.[8] As a result of various observations I thought in 1961[14]
that some gaseous effusion leaving an enduring deposit must be
involved. Patrick Moore (1962)[27] reached a similar conclusion.

For the time being, however, we have no complete answer to
the nature and origin of crater rays. Like the craters themselves,
they need not have been produced by one and the same process
in every case and several distinct causes may contribute to the total
effect in one and the same aureole. The close-up photographs are
not unambiguous in this respect, and even the proposed landing in
Mare Tranquillitatis seems unlikely to bring a final decision
between the rival conceptions.

REFERENCES

1 ARTHUR, D. W. G. (1954). 'The Classification of Smaller Lunar Craters', *J.B.A.A.*, **64** (1).

2 BEALS, C. S. (1958). 'A Survey of Terrestrial Craters', *Nature*, **181**, p. 559.

3 CRAWFORD, A. R. (1963). 'Large Ring Structures in a South Australian Volcanic Complex', *Nature*, **197** (4863), p. 140 ff.

4 DAUVILLIER, A. (1955. *Cosmologie et Chimie*. Presses universitaires de France, Paris.

5 DE SITTER, L. U. (1956). *Structural Geology*. McGraw Hill, London.

6 ELSTON, W. E. (1965). 'Rhyolite ash-flow plateaus, ring-dike complexes, calderas, lopoliths, and Moon craters', *Annals of the New York Academy of Sciences*, **123**, pp. 817–842.

7 ELSTON, W. E. (1965). 'Possible shatter cones in a volcanic vent near Albuquerque, New Mexico', *Annals of the New York Academy of Sciences*, **123**, p. 1,003 ff.

8 FIELDER, GILBERT (1961). *Structure of the Moon's Surface*. Pergamon Press, Oxford, Etc.

9 FIELDER, GILBERT (1965). *Lunar Geology*. Lutterworth Press, London.

10 FIELDER, GILBERT (1966). 'Tests of randomness in the distribution of lunar craters', *M.N.R.A.S.*, **132**, p. 413 ff.

11 FIELDER, GILBERT (1967). 'Volcanic Rings on the Moon', *Nature*, **213** (5074), p. 611 ff.

12 FIELDER, G., and WARNER, B. (1962). 'Stress systems in the vicinity of lunar craters', *Planetary and Space Science*, **9**, pp. 11–18.

13 FIRSOFF, V. A. (1959). *Strange World of the Moon*. Hutchinson, London.

14 FIRSOFF, V. A. (1961). *Surface of the Moon*. Hutchinson, London.

15 FIRSOFF, V. A. (1963). 'Selenological Implications of the South-Australian Ring Structures', *Nature*, **198** (4875), p. 78 f.

16 FIRSOFF, V. A. (1965). 'A Preliminary Selenological Analysis of *Ranger 9* Photographs'. Preprints sent to the NASA agencies in Washington and Goddard, published in 1967, *J.B.A.A.*, **77** (2), p. 106 ff.

17 FIRSOFF, V. A. (1967). 'Close-up Views of the Lunar Ground', *J.B.A.A.*, **77** (4), p. 251 ff.

18 FIRSOFF, V. A. (1968). 'Water within and upon the Moon', *New Scientist*, **37** (587), p. 527 ff.

19 GOLD, THOMAS (1955). 'The Lunar Surface', *M.N.R.A.S.*, **115** (6).

20 GOLD, THOMAS (1966). 'The Moon's Surface', *Proceedings of the 1965 IAU–NASA Symposium*, p. 99 ff. The Johns Hopkins Press, Baltimore.

21 HOLMES, ARTHUR (1965). *Principles of Physical Geology*. Nelson, London and Edinburgh.

22 JEFFREYS, SIR HAROLD (1952). *The Earth*. 3rd Edn. Cambridge University Press.

23 KATTERFELD, G. N. (1967). 'Types, Ages and Origins of Lunar Ring Structures', *Icarus,* **6** (3), p. 360 ff.

24 KOPAL, ZDENEK (1968). *Exploration of the Moon by Spaceship.* Oliver & Boyd, Edinburgh and London.

25 KUIPER, G. P. (1966). 'The Surface Structure of the Moon', *Proceedings of the 1965 IAU–NASA Symposium,* p. 99 ff. The Johns Hopkins Press, Baltimore.

26 LOWMAN, O. D., JR. (1963). 'The Relation of Tectites to Lunar Igneous Activity', *Icarus,* **2** (1), p. 35 ff.

27 MOORE, PATRICK (1962). 'The Origin of the Lunar Rays', *J.B.A.A.,* **72** (5).

28 MOORE P., and CATTERMOLE, P. (1967). *The Craters of the Moon.* Lutterworth Press, London.

29 O'KEEFE, J. A., and CAMERON, WINIFRED S. (1962). 'Evidence from the Moon's Surface Features for the Production of Lunar Granites', *Icarus,* **1** (3).

30 O'KEEFE, J. A. (1966). 'Lunar Ash Flows', *Proceedings of the 1965 IAU–NASA Symposium,* p. 259 ff.

31 O'KEEFE, J. A., LOWMAN, O. D., JR., and CAMERON, W. S. (1967). 'Lunar Ring Dikes from Lunar Orbiter I', *Science,* **155** (3758), pp. 77–79.

32 PICKERING, W. H. (1903). *The Moon.* Doubleday, New York.

33 PIKE, R. J. (1964). 'Some Morphometric Properties of the Lunar Surface', *Cornell Aeronautical Laboratory Report No. VS–1985–C–1.*

34 SATO, TAKESHI (1968). Private communication.

35 SCHATZMAN, EVERY (1966). *The Origin and Evolution of the Universe,* p. 19. Hutchinson, London.

36 SINGER, S. F. (1957). 'The Origin and Age of Meteorites', *Irish Astronomical Journal,* **4** (6).

37 SPURR, J. E. (1944–1949). *Geology Applied to Selenology,* I–IV. Science Press, Rumford Press and Business Press.

38 STROM, R. G., and FIELDER, G. (1967). 'Multiphase Development of the Lunar Crater Tycho', *Nature,* **217** (5129), p. 611 ff.

39 TYRRELL, W. G. (1931). *Volcanoes.* Thornton Butterworth, London.

40 WARNER, BRIAN (1961). 'Inner Rings in Lunar Craters', *J.B.A.A.,* **71** (3), p. 115 ff.

41 WILLIAMS, HOWEL (1941). 'Calderas and Their Origin', *Bulletin of the Department of Geological Sciences. University of California Publications,* **25**, pp. 239–246.

Abbreviations used

J.B.A.A.　=Journal of the British Astronomical Association.
M.N.R.A.S.=Monthly Notices of the Royal Astronomical Society.

11

Rivers? — impossible!

It is supposed to be a hallmark of scientific 'objectivity' to regard any suggestion of water on other worlds and the Moon in particular with intense scepticism, approaching pious horror. Yet in sober truth hydrogen is the most abundant element in the universe, and, if we disregard the chemically inert helium and neon, oxygen comes next, so that water must inevitably be the commonest of all compounds, and not its presence, but its absence should be a matter for comment.

The problem of lunar waters has already been touched upon in Chapters 5, 9 and 10. We have seen in the latter that there is good morphological evidence for at least some of the small craters having been formed by phreatic eruption (p. 137). Kurd von Bülow in his 'Beiträge zur Selenologie' (1964)[1] goes so far as to suggest that a feature of the dimensions of Sinus Iridum may be due to the same agency. This seems to be going a little too far, and v. Bülow himself would have hardly resorted to this explanation if he realized that the true outline of the 'Bay of Rainbows' is not arcuate, but a part of a hexagon, conforming to the general lineaments of the grid system. In ordinary telescopic views the polygonal shape of the huge bay, 250 km wide, is disguised by the great curvature of the lunar globe, but it stands out clearly in spherical projection photographs where Sinus Iridum is shifted nearer to the limb (Firsoff, 1961).[6] For the same reason it cannot be the 'impact channel' of the asteroid supposedly imbedded somewhere beneath Mare Imbrium. It is quite obviously a collapse feature, intensified by an isostatic uplift of the Jura Mountains – a point to be considered further in the next chapter.

On the other hand, I would endorse v. Bülow's analogy between such lunar craters and what he describes as 'pseudo-craters' blown-out by superheated steam in the Myvatn district of northern Iceland when lava overflowed some water-soaked ground. Also valid is the point made by him that a phreatic eruption involves a reduction in the volume of the material beneath the crater floor, which may, therefore, be expected to lie below the level of the surrounding country. It may also result in a substantially unwalled subsidence of stepped configuration, exemplified by Námaskard, east of Myvatn.

A phreatic eruption, however, is an explosion, although its violence may be graded according to the temperature of the superheated water at the moment of its conversion to steam, and on the Moon this temperature will generally be low. Yet there is a class of bowl-shaped craters, mentioned on p. 67, which are so perfectly formed 'as though they had been turned out on a lathe' (I think this remark is due to G. P. Kuiper). A further interesting feature of these craters is that their rim usually conforms to the horizontal plane, disregarding the slope of the ground.

Quite a few of them will be seen in the *Orbiter II* views of Copernicus or the *Orbiter III* photograph of Kepler.

There do not seem to be any ejecta, despite the recent appearance of the depression, and such boulders as may occur near the lip of the crater are attributable to erosion. Nor are there any indications of marginal sloping, characteristic of cauldron subsidence. This would exclude impact, phreatic or ordinary volcanism alike as a possible explanation.

At the same time these bowl-shaped craters occur in association with equally regular round domes of comparable dimensions, and it is natural to infer a connection between the two. The craters would, then, be 'deflated' domes. In order to produce the desired effect, the generative process must be slow and gentle. Pingo is the only terrestrial form that appears to meet the requirements (p. 109).

Terrestrial pingos are found in recently glaciated terrain, sometimes over frozen lakes, and the direct analogy may be misleading. Typically a pingo is formed where 'unfrozen ground water (a) is trapped between an underlying zone of permafrost and an overlying frozen zone, or (b) has access to such a position from a high intake area, which gives it a hydraulic head like that of artesian water' (Holmes, *Principles of Physical Geology*, p. 432).[9] In freezing the intruded water experiences a 9 p.c. expansion and lifts up a dome. Some Canadian pingos are up to 140 feet (say, 45 metres) high. If the soil covering at the top of the dome is breached, some of the ice melts, causing a partial collapse into a caldera, usually cradling a lake.

The lunar situation is comparable, but not equivalent, the only direct resemblance being the existence on the Moon of the permafrost seal, which the rising water encounters on the way to the surface. If the seal is broken the most likely result is a phreatic eruption. But liquid water seeping into the permafrost layer would uparch it little by little, freezing up at the same time. A liquid cannot be porous, but under the low lunar gravity ice may be expected to contain bubbles of steam and assume a foam-like, expanded texture. Thus the gain in volume at freezing may greatly exceed the terrestrial 9 p.c. The rock overburden is also lighter and can be lifted up much more easily. To sum up, a much larger and steeper *cryo-laccolithic* dome should result.

The main difference between the Earth and the Moon, however, stems from the fact that liquid water cannot appear unless the barometric pressure rises over 4·58 mmHg. This cannot be normally expected in the present-day conditions of the lunar surface. Below this triple point ice boils without melting, or *sublimes*. O'Keefe, Lowman and Cameron argue against pingos (1967),[13] at least in the equatorial zone of the Moon, but we may take their figures regardless of the argument.

If we put the subsoil permafrost temperature at 240° K from microwave data, the corresponding vapour pressure of ice is 0·3 mmHg. Assuming, further, sublimation into vacuum, they give the rate of evaporation as 1 g/cm²/min. They deem this impossible, because – they say – 'at this rate, the loss from an area 100 m in diameter would in one day equal the whole lunar atmosphere'. For this argument to carry any weight we must be sure not only of the present, but of the past mass of the lunar atmosphere, and, as we shall see later on, the latter may have been appreciable. Nor is even the present situation as clear as is often assumed. This also means that sublimation to vacuum may be an exaggeration.

What is really important, however, is that in a lunar pingo the ice will not melt superficially, as it does on the Earth, but fizz away completely, right to the bottom, fairly rapidly, yet peaceably. This seems to provide an ideal mechanism of formation to explain those 'lathe-made' bowl-craters.

It may here be added that some of the cratered domes on the inner glacis of Copernicus bear a strong resemblance to terrestrial pingos, which, though, could be a very good reason for their being something else.

Urey[16] and Gold's idea that the maria may be underlain by a layer of plastic ice and its rejection by O'Keefe[12] on the ground of the low viscosity of ice, which would cause all major features of surface relief to collapse within years, if not months, has already been considered on p. 110. It may be repeated that O'Keefe's criticism is valid in this context. Nor is it perhaps particularly likely that Alpetragius is a pingo. Nevertheless, the contrary assurance smacks of prejudice, for water, whether frozen or liquid, can behave like incompetent rock and simulate igneous tectonics. If the maria were similarly based on a layer of 'plastic magma' O'Keefe's argument would hold, as the viscosity of magma is still lower than that of ice; but it is self-evidently absurd to deny for this reason the reality of igneous intrusions and their consequences, such as collapse calderas. Only local ice and water action is envisaged here.

O'Keefe, Lowman and Cameron[13] also inveigh against the subsurface ice flow in the manner of the so-called 'rock glaciers' of Alaska. The latter have been investigated by Wahrfahtig and Cox, who have found that they were not just rivers of dry rock rubble, but the rubble was riding on solidly buried glacier ice. According to O'Keefe *et al.* this is impossible on the Moon, as the ice could not be effectively sealed from contact with the sub-vacuum of the surface and would soon sublime away. In the immediate context of this argument, which is the nature of the peripheral heights of the Flamsteed Ring, I fully agree that these show no indication of being anything but extrusions of highly-viscous lava from a ring fracture – certainly not 'rock glaciers'. It may also be conceded that rock rubble does not constitute an effective seal for glacier ice in sub-vacuum conditions, but there is a great difference between the merely improbable and the absolutely impossible.

We will recall here the curious aprons on the north-western

glacis of the great mountain ring of Tsiolkovsky (p. 84). What are they?

Lava flows, whether terrestrial or lunar (cf. *L.O. IV* frame 173, 2 of 3), are characterized by transverse ribbing, like frozen waves, which in a sense they are. Ash flows may sweep downslope along straight or gently curved, interlacing lines, as may be seen to advantage on the outer glacis of Aristarchus (Plate 11a), and run out into a thin feathery overlay. Landslides, being gravitational movements of incoherent material, tend to be structureless. On the Moon, for instance, the crater Clarkson sends what seems to be a landslide into Gassendi. The *L.O. V* frame HR 129, 1 of 3, shows two overlapping craters with a landslide at the overlap – though this could perhaps be lava. A mud flow, or *lahar*, is a somewhat unlikely development on the Moon, as it would be immediately transformed into an ash flow through the vaporization of its water content. There appears, however, to have been a period in the Moon's history when it had an atmosphere sufficient for the free flow of water on the surface, so that we cannot afford to be dogmatic about that either. In any case a lahar is morphologically comparable to a flow of lava, and orbital photography offers no sure means of distinguishing between the two.

Yet the big apron of Tsiolkovsky does not conform to any of these patterns. There are in it a few transverse waves or breaks, but its most distinctive feature is a strong longitudinal striation, slightly divergent towards the foot and clearly defining lines of flow. This could in itself fit an ash flow explanation, but the apron does not 'feather out' and ends up abruptly in a steep front, not less than 100 metres high. Quite obviously it is a body of considerable mass and thickness. The appearance of its head, where it is separated from the wall of Tsiolkovsky by a deep and broad ditch, reminiscent of a glacial *bergschrund*, serves to confirm this conclusion. The ditch continues along the south-east edge of the apron, indicating both partial retreat and excavation (Plate 14).

Somewhat similar striations occur elsewhere on the glacis, which are generally free from aprons, although there is another smaller and not so well defined apron to the north-east of the main one. Thus a suspicion is aroused that these striations may be due to other aprons, which have since disappeared. Lava flows or landslides do not vanish, but a glacier may do so, either through melting or evaporation, more or less completely, leaving behind only the grooves worn out by it and a deposit of morainic material. In fact, the slopes of some ring mountains in this region display peculiar flat excavations bounded by mounds that could be terminal moraines. But the aprons are lacking. The only possible other candidate is at the edge of a minor crater facing Tsiolkovsky across an old thalassoid invaded by it. This apron, however, is so small and narrow that its nature is not at all clear.

The continuity of the striae suggest a slow movement of massive material, or rheid,[9] such as ice. Moreover, the same structure reappears in the aerial views of some terrestrial glaciers. Salt glaciers are possible, but there is other evidence of water action on the Moon, so that chances are strong that we have here a proper piedmont glacier, after all. In order to persist, it must be thoroughly shielded by a thick and continuous overlay of some kind, and the overlay seems to be of volcanic ash, which fits the underlying surface like a glove, and, especially if periodically renewed, could provide a very effective seal.

The main witness to the past water action on the Moon, however, is borne by the sinuous rilles (already described on p. 75 f.), which their discoverer Pickering thought of as, and christened, 'riverbeds'.[14] Yet only the largest of these tortuous channels can be seen through a telescope at all and indistinctly at that. Thus Pickering's idea could be easily waved aside as one more irresponsible fancy, until the *Orbiter* photographs made these features accessible to closer study (Plates 15, 16, 17).

It will be recalled here that typically a sinuous rille originates in an elongated, often pear-shaped, crater on relatively high ground, such as the slopes of a large mountain ring, and pursues its downward course along the natural declivities, avoiding all elevations. This course is sinuous, as the name implies, and meanders intricately when the angle of slope is small or obstructions are numerous. At the same time the channel grows thinner and thiner, or at least shallower and shallower, and eventually fades away.

This is the behaviour of a stream of water in conditions of extreme dryness and high evaporation. As Pickering[14] himself had said, it is as though a lake were draining into a river, the river into a brook and the brook into a rille. The resemblance to the terrestrial river beds seen from the air is, indeed, very close, except that with high rainfall these have dendritic appendages where minor streams join the main flow.

All these characteristics have been confirmed by R. E. Eggleton and H. J. Moore of the U.S. Geological Survey, who have examined about twenty-five sinuous rilles. Nevertheless, Winifred S. Cameron (1964)[2] seeks to explain these features as channels excavated by ash flows, and instances the Kambara Ditch, which is nothing like a sinuous rille, on the slopes of the Asama volcano in Japan as a possible terrestrial example. Hers is, of course, a pre-*Orbiter* opinion, based on telescopic evidence alone, and as such may be considered obsolete. But it has been re-affirmed more recently (1967) with reference to and in the teeth of the *Orbiter* photographs.[8] The lava ditch in the Lanzarote Mountains may be a more convincing analogue of at least some lunar sinuous rilles. Having examined, however, every available *Orbiter* frame with sinuous rilles, I am convinced that they are channels worn out by freely flowing water, and no other explanation will wash.

The opposition to this is based on two arguments.

Firstly, as already stated (p. 144), the triple point of water is defined by the barometric pressure of 4·58 mmHg and 0° C. True, solutions of salts or other substances will generally freeze at lower and boil at higher temperatures. Water in small droplets can be supercooled in the absence of condensation nuclei to −72° C without freezing.[10, 11] Yet all this requires its having been liquid to begin with, and so an atmospheric pressure of over 4.58 mmHg. Even then the appearance of the liquid phase at such low pressures, especially when combined with the high midday temperatures of the Moon, would be evanescent, and a good deal more than 4·58 mmHg would be needed to allow a free flow sufficient to excavate a channel 200 km long, such as Schroeter's Valley (p. 76). An air mass amounting to 1/6 of ours would yield under lunar gravity a ground-level pressure of barely some 20 mb, which corresponds to a boiling point of +22° C; whereas the present-day noon temperatures in the equatorial zone of the Moon may exceed +100° C, and the search for a lunar atmosphere has given negative, or at best ambiguous, results.

On the other hand, in 1960 J. J. Gilvarry[8] made a calculation, based on the assumption that the initial water content of the lunar and terrestrial magmas was the same, to show that the Moon ought once to have possessed a copious atmosphere and a hydrosphere covering the present marial basins to an average depth of 2 km, and that this hydrosphere and atmosphere should have endured for at least 1,000 million years before being eventually lost to space by molecular evaporation.

I would not go with Gilvarry all the way. I think his reasoning suffers from the usual defect of paying insufficient attention to the consequences of reduced mass for the structure and development of the lunar globe. As indicated in Chapter 5, in lunar conditions the volatiles, including water, cannot be expected to have been squeezed out of the interior to the surface as thoroughly as in the more massive Earth. In other words, the Moon may have never had a hydrosphere inferred by Gilvarry. This, though, does not necessarily mean that it never had any hydrosphere at all. His thinking is scientifically sound within the assumptions made at the start, and his conclusions cannot be lightly dismissed. It is, there-fore, very unwise to be dogmatic about the lunar atmosphere in the remote, or even recent, selenological past.

The second objection to interpreting the sinuous rilles as 'river-beds' stems indirectly from H. C. Urey (1967),[16] who is puzzled by the absence of what he calls a 'delta' at the foot of the rille, where the materials eroded by the stream would be deposited. This leads him to the assumption that the channel has been cut in the ice underlying the mare surface. For one thing, however, sinuous rilles, although most conspicuous in the maria, are not wholly con-fined to lunabase. At least one rille rises in the lunarite uplands west of the crater Bode, and it is a question of interpretation whether it reaches the lunabase lowlands at all; there are several sinuous rilles in the Jura mountains. For another, the conception of ice-based maria has been effectively demolished by O'Keefe[12] (p. 110).

Once again, however, the weakness of this reasoning lies in disregarding the effects of the low lunar gravity and thinking in purely terrestrial terms. If we put the mean bulk density of the sur-face formations at 0·2 (p. 58), and consider that in being eroded by running water they will lose their foam structure and must be compacted to a much higher bulk density, say, between 1 and 2, we shall see that the thickness of the resulting deposit must be very small in relation to the eroded volume. It could thus be wholly absorbed within the bed itself. Nor is it true that there are no signs of deposition.

Consider the Hadley and Conon Rilles, descending from the Apennines into Palus Putredinis and Mare Vaporum respectively. Both channels are trenched very deep. Neither grows any thinner as it progresses, but the former runs very shallow after rounding the long horst north-west of the Apennines, where it can be made

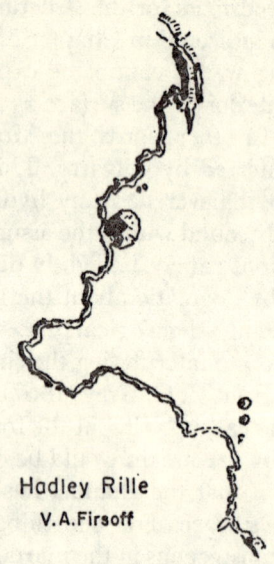

Hadley Rille

V. A. Firsoff

Fig. 11. The Hadley Rille, one of Pickering's original 'riverbeds', descends from Mt. Hadley in the Lunar Apennines into the Sea of Rains (Mare Imbrium) in a deep clear-cut channel. Drawn by the author from an *Orbiter V* photograph. (See also Plate 2).

out only with difficulty (Fig. 11). This is partly due to its coursing parallel to the direction of sunlight, but the greatly reduced depth is undeniable and must be ascribed to silting, for it is here that the rate of flow may be expected to slacken. The subsequent recovery and increased depth towards the end is probably caused by a fresh influx of water from the neighbouring hills. The Conon Rille may be vaguely traceable upwards into the highlands about the crater of that ilk, and it does receive some headstreams from the north. Its clear origins, however, are in a complex of shallow 'lakes' at the head of the lunabase valley, where it achieves a considerable depth without any preamble, then to stop as abruptly after not more than 30 miles. Under favourable illumination a 'ghost stream' appears to go on, but it has obviously been blocked solid by the material brought from above.

The argument from the absence of a delta is not valid.

To my mind, however, the issue is clinched not by the largest and most conspicuous 'riverbeds' known from telescopic observation, but by the tiniest specimens revealed by the high-resolution orbital photography. The *L.O. IV* frame HR 187, 2 of 3, gives a beautiful view of the north-east corner of Mare Orientale and the adjoining lunabase valleys of Maria, Hiemis and Autumni. In the latter two thin sinuous rilles can be made out, preferably with a magnifying glass, as they are close to the threshold of resolution. The more interesting of the two rilles is above the centre line on the right. It can be traced through the lunarite* some way up into the Rook Mountains, but its exact origin is not easy to pinpoint. Having reached the level floor of the Mare, it forms innumerable meanders, skirts a small hill and continues more briskly southwards to and through a narrow gap between the rocky shores, dying out a couple of miles farther on. It is simply not credible that a feature like this could have been formed by anything other than flowing water.

There are similar small streams in Mare Imbrium at the foot of the Apennines near Eratosthenes, and even thinner meanders that remain unresolved may be suspected.

The sinuous rilles are associated with the maria, and most of them are to be found in and around Mare Imbrium and in Oceanus Procellarum, the Lunar Alps and the region extending from Sinus Roris in the north through Aristarchus to Marius being particularly prolific. They seem to be substantially absent from the averted hemisphere, with the notable exception of 'Object A' described in the next chapter (p. 166). Some channel-like lines on the inner slopes of Tsiolkovsky and near that great ring could also be due to water action. Indeed, sinuous rilles are always difficult to follow into lunarite. The whole area of the Apennines near

---

* This effectively disposes of the, lava-ditch, idea as a lava-ditch could not continue from the lunarite into lunabase.

Palus Putredinis and Cape Fresnel gives an impression of superficial moulding by water, especially in the *Orbiter* photographs of
medium resolution. One or two streams seem to streak down Cape
Fresnel itself from the crater in the south. It will also be appreciated that tectonic fractures on the Earth readily become natural
drainage channels, and the same must be expected on the Moon.
For instance, the trough-like feature west of Cape Fresnel seems
once to have cradled a very substantial lake. This is also true of the
graben-riven country south of Palus Putredinis, and it is not
beyond belief that Whitaker's 'Lake Titicaca' on the west rampart of Alphonsus was once filled with water. If so, however,
some crater bowls, too, could have held water at that time, as they
frequently do on the Earth. More particularly the shallow irregular depressions, occasionally found in marginal lunabase regions
and suggestive of Ugandan maars, look as though they were once
filled with water. This suspicion hardens into certainty where they
are actually connected with one or more sinuous rilles. Such old
lake beds are clearly visible near the head of the Conon Rille.

Far from being a picture of mournful desolation, as it appeared
to Borman, Lovell and Anders through the windows of the *Apollo
VIII*, the Moon at that time would have been a beautiful place to
the human eye. Look at the Alps in Plate 16. Imagine their craters
gleaming with lakes and water coursing down those channels
weaving away to Mare Imbrium over broad straths between noble
mountains wreathed with clouds! Was there life in those valleys
and on the now-barren hillsides? Even the landing in Mare Tranquillitatis could not answer this question in the negative, although
any evidence of past life found there may be regarded as a
resounding 'Yes' for the Lunar Alps as well. In fact, here and there
the mountain-sides display peculiar concentric curves, possibly
marks of lava flow, but they do suggest the remains of some
growths, as though of gigantic lichens . . . And if there was life
on the Moon's surface at any time in the past, it may still exist
underground – in those 'caverns measureless to man'.

We may here return to the Bode Rille. It is born in a well-
behaved way from an elongated crater, or rather crater-chain, on
the western slopes of Bode and flows north, with many twists and
turns, for a space of about 50 km, to plunge over the edge of the
hills into a broad roughly rectangular, flat-bottomed depression,
some 40 by 10 km, which shows inner shore lines and must have

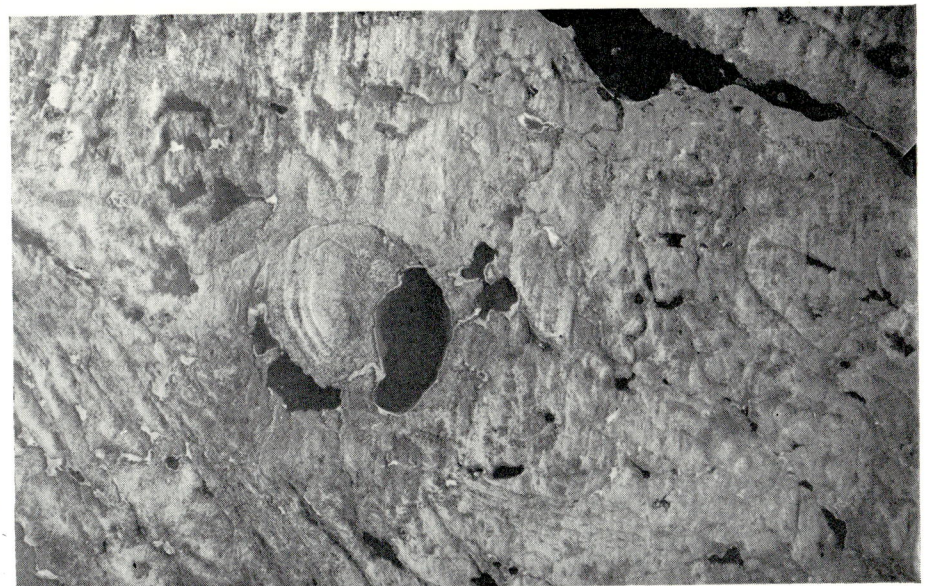

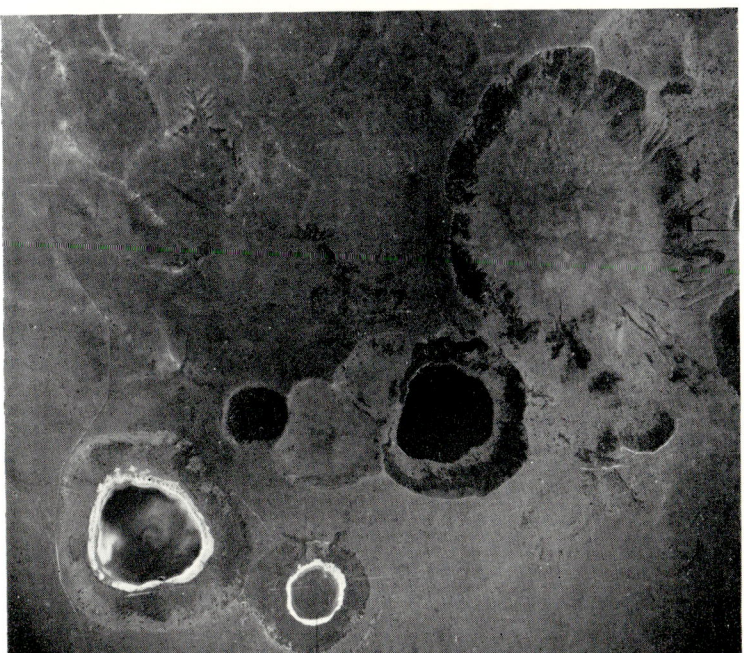

Plate 12a. The Brent Meteoric Crater, reckoned at 500 million years, is 3.4 km. in diameter. Note also the striated appearance of the country due to recent glaciation. *The Royal Canadian Air Force.*

Plate 12b. Volcanic maar craters, outlined by water and vegetation, in the Queen Elizabeth National Park, Uganda. *Department of Lands and Surveys, Entebbe, Uganda.*

(a)

(b)

Plate 13a. Lava flows near Inghirami (large ring at the top). A part of the infilled ring of Wargentin can be seen in the bottom right corner (IV–180H₁).

N.A.S.A.

Plate 13b. 'Object A' (p. 166), a large walled plain that may once have cradled a lake, photographed by *Orbiter II* in the southern part of the averted hemisphere.

U.S. Information Service, London.

Plate 14. The great ring of Tsiolkovsky on the Moon's far side (III–121M₁). *N.A.S.A.*

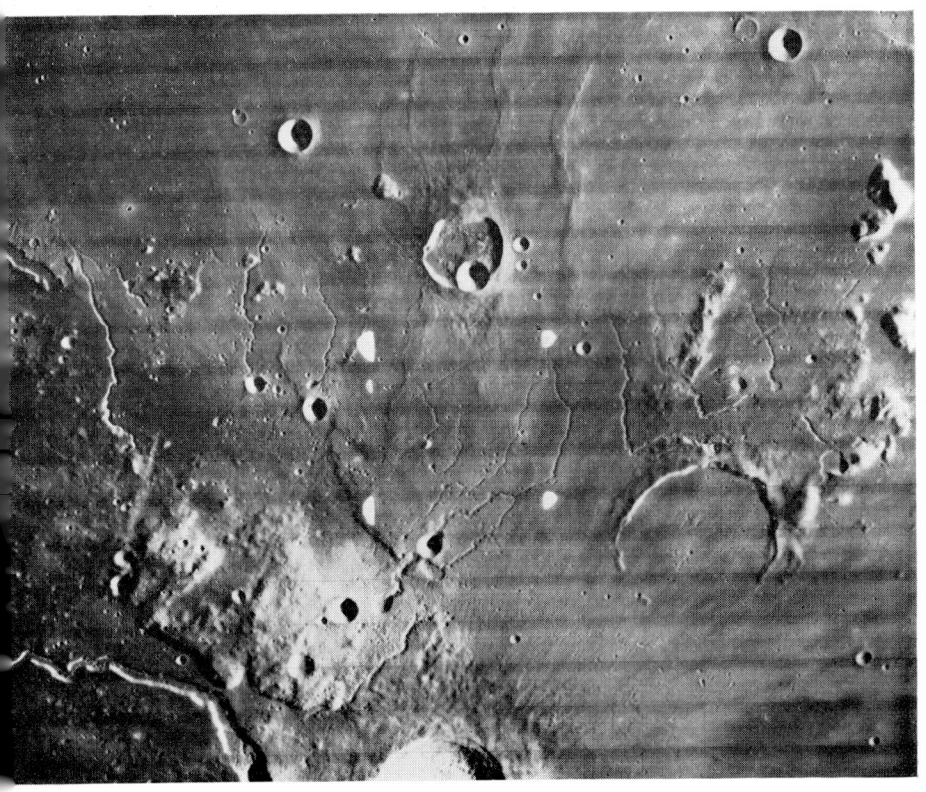

Plate 15. Sinuous rilles in the region of Aristarchus and Prinz, Schroeter's Valley is in the bottom left corner [only about a third of its length is shown] (IV–151H₁).

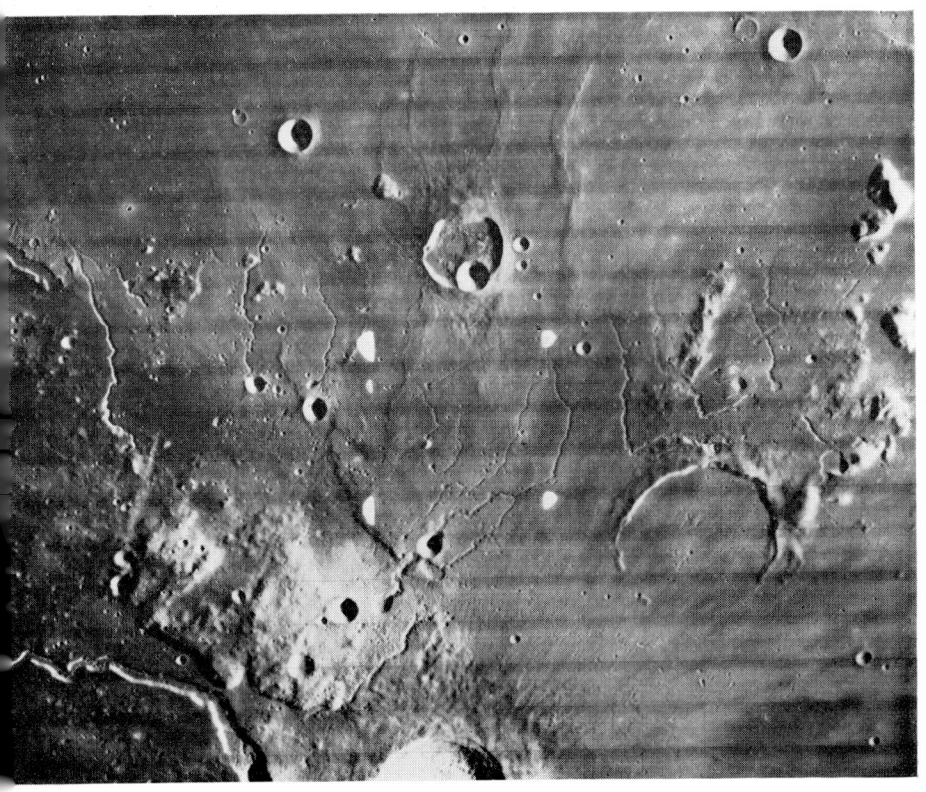N.A.S.A.

Plate 16. A medium-resolution *Orbiter-V* view of the Lunar Alps with the Great Alpine Valley and a system of river-beds (V–131M₁).

N.A.S.A.

Plate 17. Gassendi and Clarkson rings, parts of Mare Humorum and Oceanus Procellarum with sinuous rilles, one of which cuts through a range of hills (p. 155) (IV–137H₂).

N.A.S.A.

been a lake for a time, comparable to, say, Lago di Garda. The rille reappears at the north end of the depression, and continues to meander down a flattish dark upland for another 50 km or so towards Sinus Aestuum, although it is not quite clear if it has ever reached the latter or not. It seems, however, to have received some tributaries from the south, and another thin stream, rising in a small crater-like hollow in the north of the plateau, may have joined it from the other side.

Why are the hills here so dark? This may, of course, be simply due to the same kind of dark deposit as is found round some craterlets in Alphonsus; but Gilvarry sought to ascribe the dark tint of the maria to carbon derived from the organisms that used to live there when waters ebbed and flowed in the powerful tides of the Earth between the now desolate shores. I would not say I was convinced by his arguments on this point. But who knows? – perhaps he was right, after all. The answer may soon be forthcoming.

The sinuous rilles are not all alike. They do not inevitably issue from a crater. Some are comparatively broad and shallow, flat-floored ravines, suggestive of a copious but short-lived flow acting on a readily erodible material. A few of the rilles about Prinz may be put in this category; but probably the best case in point is the long ghostly ribbon meandering into Mare Imbrium past Wallace from the foothills of the Apennines (*L.O. IV* HR 114, 1 of 3). It is so shallow that it can be descried only intermittently in the late morning light, which brings out quite clearly the slender channels farther west.

On the other hand, the main river of the Alps, between the 'Great Valley' and Plato is doubly trenched, especially in its middle course, a V-shaped channel cutting into the floor of a wider ravine, which indicates either a gradual dwindling of the water supply, or its seasonal fluctuation. The banks are thinly coated with some loose material, such as volcanic ash, to a variable depth. But farther north an old sinuous rille has been completely filled with it. Together with the volume of erosion, this goes to show that the 'water episode' extended, if perhaps spasmodically, over a considerable time. The impressive depth of the Hadley Rille points in the same direction.

The rille to the north-east of Plato is cut even deeper over its short course. Although not very sinuous, it emerges properly from an elongated crater, and is certainly a river bed. It disappears below the ground just north of a group of peaks, seemingly swallowed by the crustal fracture from which these have been extruded, but a number of smaller rilles pearl out of the hills there and probably communicate with the main channel. Special interest attaches to the easternmost of these which continues the line of a crater chain. It may simply be that the spring was hot, and the superheated water, flowing underground, blew out holes in the roofing rocks; or else the atmospheric pressure may have been temporarily too low for liquid water. Yet our mountain streams not infrequently vanish beneath boulder screes or into limestone caverns, to bob out again farther on.

The largest, if not the longest member of the whole sinuous tribe, however, is Schroeter's Valley, born from the Cobra-Head, an elongated depression on the northern slopes of Herodotus, to zig-zag away over 200 km or so into Oceanus Procellarum. It is up to 4 km wide at the head, but even its central part is not less than 2 km from bank to bank, as well as some 300 metres deep. The sidings are pretty steep, too, and may exceed 60° in places, abutting quite sharply at a nearly level floor, over which a thinner, dead watercourse threads its way, now along the one, now along the other wall. This would point to a depression at the foot of the walls, which assimilates the structure of the canyon to that of graben-rilles (p. 75). Its course, too, is far more angular than that of a typical sinuous rille, and suggests a tectonic collapse (Plate 15). Possibly we have here something comparable to our own Cheddar Gorge, where the limestone roof has caved in over an underground river, as I suggested in 1959[5] (Gold has recently proposed the same explanation). There are, however, other similar canyons in the Jura highlands and on the south-west slopes of Plato which could be due to simple erosion.

Be this as it may, Schroeter's Valley is not all that different from the Hyginus Cleft, usually regarded as a graben-rille. The latter is shallower, and there is no river bed on its floor, but the craters that are strung along it may be phreatic; water on the Moon is clearly associated with volcanic action, nor is there any good reason why tectonic movements could not contribute to the total effect.

The eastern leg of the Hyginus 'Cleft' resembles another peculiar rille. It is one of the Marius complex in Oceanus Procellarum, and its close-up view appears in the *L.O. V* frame HR 213, 2 of 3. It has the 'regulation' tadpole contour and is classed among the sinuous rilles. Yet it does not sinuate much, and, moreover, it cuts through a wrinkle ridge. It may well be that the ridge is seleno-logically younger than the rille (a long winding rille near Gassendi also crosses a minor range of hills which would appear to have been elevated while the river was flowing, thus bearing witness to the length of time over which it existed – Plate 17), but the general appearance of the latter is unusual. Its cross-section oscillates between a V and a U profile, and the whole gives the impression of having been gouged with a blunt tool.

This invites reflection.

The lunar day being equal to about 29 days and 13 hours of our reckoning, the nights are very cold. In the present airless times the thermometer may sink to 150° C of frost. An atmospheric envelope would temper such extremes, both by direct shielding and circulation; even so severe night frost seems inevitable. It could, of course, be that the Moon rotated much faster in the 'water period', but this seems unlikely because the sinuous rilles are among the most recent features of the Moon : not only are they obviously post-mare, but they are posterior even to such post-mare rayed craters as Aristarchus, to say nothing of Mare Orien-tale, which, too, must be reckoned one of the youngest – perhaps the youngest of all – major lunar formations.[7] By then the lunar orbital regime could not have differed greatly from the present.

This means, however, that all watercourses would freeze at night; they might, in fact, freeze right through to the bottom. If now the volume of liquid were large, or it continued to rise to the surface, a glacier would result. The shape and behaviour of this Marius rille could then be accounted for by a combination of ice and water erosion.

J. E. Spurr was not a believer in a dense lunar atmosphere at any selenological period, but he did envisage massive volcanic eruptions, involving the emission of great amounts of gas, which were chiefly steam, and generating a transitory local atmosphere, or *atmodome,* as he called it. Condensing within the atmodome, water vapour would release brief torrential downpours, which scoured the sides of the 'lunavo' (p. 108) with radial gullies, such as may be seen around Timocharis, Aristillus and some other marial craters.[15] Fielder and Warner's investigation appears to show that these are tectonic in origin,[4] but thin sinuous rilles meander across the 'lava lakes' on the slopes of Tycho; and here and there rapid torrents seem to have washed down the inclines of this crater as well as of Copernicus and Aristarchus (Plate 11a).

Seeing that Aristarchus is particularly associated with sinuous rilles, there is no need to quibble over this interpretation. Furthermore, if we take, for instance, the *L.O.V* frame MR 202, 1 of 1, we shall observe that some of the ash at the lip of the crater appears to have been washed away.

To sum up, it is no longer possible to exclude at least local water and ice action on the lunar surface, and, even if the aprons of Tsiolkovsky are not actually extant 'rock glaciers', some, perhaps extensive, glaciation must be considered a serious selenological possibility.

REFERENCES

1   BÜLOW, KURD VON (1964). 'Beiträge zur Selenologie, IV', *Geologie*, **13** (6/7), p. 899 ff.

2   CAMERON, WINIFRED S. (1964). 'An Interpretation of Schröter's Valley and Other Lunar Sinuous Rilles', *Journal of Geophysical Research*, **69** (12), p. 2,423 ff.

3   CAMERON, WINIFRED S. (1967). 'Vulcanism on the Moon', *Science and Children*, **5** (4).

4   FIELDER, G., and WARNER, B. (1962). 'Stress systems in the vicinity of lunar craters', *Planetary and Space Science*, **9**, pp. 11–18.

5   FIRSOFF, V. A. (1959). *Strange World of the Moon*. Hutchinson, London.

6   FIRSOFF, V. A. (1961). *Moon Atlas*. Hutchinson, London.

7   FIRSOFF, V. A. (1968). 'Water within and upon the Moon', *New Scientist*, **37** (587), p. 251 ff.

8   GILVARRY, J. J. (1960). 'Origin and Nature of Lunar Surface Features', *Nature*, **188** (4754), p. 886 ff.

9   HOLMES, ARTHUR (1965). *Principles of Physical Geology*. Nelson, London and Edinburgh.

10 KISTLER, S. S. (1936). 'The Measurement of "Bound" Water by the Freezing Method', *Journal of the American Chemical Society*, **58** (6).

11 MAMIKUNIAN, G., and BRIGGS, M. (Eds.) (1965). *Current Aspects of Exobiology*, Pergamon Press, Oxford.

12 O'KEEFE, J. A. (1968). *Water on the Moon and a New Non-Dimensional Number*. Preprint.

13 O'KEEFE, J. A., LOWMAN, O. D., JR., and CAMERON, W. S. (1967). *Science*, **155** (3758), p. 77 ff.

14 PICKERING, W. H. (1903). *The Moon*. Doubleday, New York.

15 SPURR, J. E. (1944). *Geology Applied to Selenology. The Imbrian Region of the Moon*. Science Press.

16 UREY, H. C. (1967). 'Water on the Moon', *Nature*, **216**, p. 1,094 f.

12

The origin and nature of the Maria

When the distance is sufficiently large in relation to the size of a mass the latter may be regarded as though it resided in its centre of gravity, but close up to an extended body the attraction of its parts becomes noticeable. On the Earth's surface gravity varies slightly from place to place on both sides of the mean value of 978 cm/s^2, not only with altitude and latitude, but by reason of differences in subsurface density, local excesses in this, known as *mascons* (mass concentrations), causing an upward departure

The flight of the *Lunar Orbiter V* was tracked from radio signals over 80 consecutive revolutions of 3 hours 11 minutes each, during which it approached the surface of the Moon to within about 100 km on the near side at latitude of 2° north. When the data were processed by computer (M. Muller and W. L. Sjögren, 1968)[23] significant accelerations were discovered over the 'circular maria', indicating the existence of mascons, 50 – 200 km in diameter, about 50 km below the surface.

The Maria, Imbrium, Serenitatis, Crisium, Nectaris and Humorum, are all underlain by mascons. There is also evidence of a strong mascon beneath Mare Orientale, and a further region of high density exists between Sinus Aestuum and Sinus Medii, just north of the centre of the Earthward hemisphere. No mascons, however, have been found under the irregular maria, which von Bülow calls 'epicontinental', Sinus Iridum or the terrae.

The meteorists, and H. C. Urey in particular, have promptly concluded that the mascons must be 'high-density planetesimals' (as explained on P. 118, a 'high-density planetesimal' is a contradiction in terms) embedded beneath the lunar surface as tangible proofs of the impact origin of the 'circular maria'. It is true that at first sight Mare Orientale strongly suggests asteroidal impact, and I used to think so myself.[13, 14] The other maria would then presumably be of the same genesis, and Mare Nectaris in particular seems to replicate the structure of Mare Orientale, modified by the passage of time. It has, however, been repeatedly emphasized that there are no proper maria on the other side of the Moon, while on this side they form a distinct pattern, which

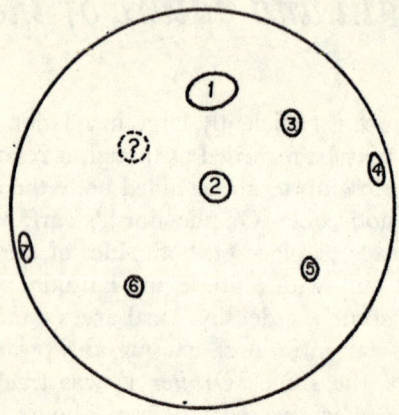

Fig. 12. The distribution of mascons on the Moon. (1) Mare Imbrium, (2) Sinus Aestuum-Sinus Medii, (3) Mare Serenitatis, (4) Mare Crisium, (5) Mare Nectaris, (6) Mare Humorum, (7) Mare Orientale. To complete the pattern, a further mascon (?) is required near the centre of Oceanus Procellarum. This is a very large lunabase area and as such ought to correspond to a positive gravity anomaly — possibly the excess of mass is less concentrated here.

comes out even more clearly in the distribution of the mascons (Fig. 12). This may be said to be centred on Mare Imbrium, as the premier 'sea' of the Moon, at a point somewhere about 35° N and 15° W, its backbone then curving southward on both sides by 30° of latitude or so, the minor maria, including Orientale and Australe hugging the libration zone. Furthermore, the western shore of Oceanus Procellarum closely follows the visible limb.

Such distribution cannot be a matter of pure coincidence and clearly relates the origin of the maria to the Earth. In fact, if we draw a great circle approximately from the assumed centre of the marial system through the *subterral point,* defined by the present centre of the near hemisphere, the pattern of the maria assumes remarkable symmetry in relation to it. Mare Nectaris corresponds to Mare Humorum, Crisium to Orientale, while the lunabase complex to the east and south-east of Mare Imbrium has its counterpart in the Nubium-Procellarum plains. If we add to this that asteroidal impacts precisely on the side facing towards the Earth are extremely improbable to begin with, and that, further, the axis of symmetry of the system coincides with one of the main trends of the pan-lunar grid lineaments, the impact hypothesis loses all plausibility.

The most probable explanation is that the maria were initiated by the body tides raised in the Moon during a close approach to the Earth at capture, when the subterral point lay in Mare Imbrium (Firsoff, 1959),[7] the axis of the lunar globe having subsequently shifted position. The great shearing stresses revealed by the eastward displacements of crustal blocks in the Alps and the Caucasus (p. 94) may then be ascribed to the arrested momentum of rotation (see also Spurr, *Geology Applied to Selenology–III*).[29] This view is corroborated by the great heat, accompanied by partial melting, which has affected the old surface in the north polar zone, but has no counterpart in the south. It may be that, as suggested by Kopal (1965),[20] the Moon grazed or entered Roche's Limit (p. 19) and its crust was shattered on the Earth's side. The only difficulty is the apparent spread in the ages of the maria. The distance in time between Mare Nectaris and Mare Orientale could hardly be less than an aeon, and the maria, as we shall see presently, offer evidence of a long history of development. They may have been initiated catastrophically, but the disturbance has taken long ages to work itself out, and we may still be witnessing its aftermath.

We must not forget, however, that the oceanic basins of the Earth, which are in many ways analogous to the lunar maria, have a preferential distribution as well. So, too, have the maria of Mars, be they depressed or elevated. Neither can be seriously attributed to impact, and at least in the latter case tidal action seems to be ruled out. In other words, internal forces – most probably mantle circulation – are by themselves adequate to produce such effects.

The forces that have presided over the birth of the maria must necessarily remain hypothetical, but the fact that at least the 'circular maria' correspond to areas of high subsurface density assimilates them to lopolithic subsidences.[18] These are isostatically induced, the crustal level being pulled down by the superior weight of the underlying magmatic intrusion, with or without a simultaneous marginal uplift. This tallies well with the ideas set forth in Chapter 5 on the partial collapse of the deep honeycomb layer and its invasion by denser material from the interior accounting both for the existence of the maria and the dynamical bulge (p. 42). The assumed depth of 50 km likewise fits this interpretation.

The lack of mascons below the terrae, irregular 'seas' and even so large a sink as Sinus Iridum suggests a basic difference between the subcircular maria and the mountain rings, the former being isostatic subsidences and the latter collapse features due to the withdrawal of magmatic support. The difference, however, may be one of degree only, isostasy intervening to some extent in the large walled plains; while marginal volcanic activity in the maria will cause the expulsion of some of the subsurface material and an ensuing coastal subsidence, with or without magmatic effusion. This appears to be the case of Mare Serenitatis, which is bordered in the east and south by a marginal trough of a darker colour. On the other hand, a partial collapse of the honeycomb structure may by itself induce subsidence without isostasy, which should be comparatively inefficient on the Moon. The collapse within the irregular, or epicontinental, maria may be of this nature. Here the densification is less and the layer of lunabase spreads thinly over the submerged crust, as is indicated, for instance, by the relative frequency of elementary and ghost rings. There also exist some semi-marial regions, such as the Hainzel-Schickard-Schiller area, which lie as low as or lower than some maria, but have escaped flooding with lunabase. They would mark a further step down in the mare-forming activity, or else the first step up in the development of a mare. Nor is the process necessarily irreversible. The post-mare rings and tholoids are lunaritic, perhaps owing to magmatic fractionation at depth, such as is involved in the production of terrestrial granites. In the course of time the lunabase may become smothered under a new layer of lunarite and a mare or its part cease to be. Mare Nectaris may once have extended to the Altai, and some ancient maria may have disappeared altogether.[6, 9]

Conjectures apart, we have in Mare Orientale (Plate 5) a recent and spatially isolated specimen of 'circular mare' available for study. The central lunabase plain is smaller than Mare Marginis or even than the hexagonal lunabase sink centred at latitude 36° S and longitude 174° E, which we shall call 'Object A' for furuther reference (Plate 13b); but the forces involved in the formation of Mare Orientale are of an altogether different order, as is amply attested by the vast area of disturbance, over 1,500 km in diameter, that surrounds it. The triple girdle of heights enclosing the basin has already received some attention on p. 000, with the result that their spacing satisfies the requirements of a tectonic depression, although this does not necessarily preclude initiation by impact. These mountain ranges, including the Cordilleras, the Rooks and the d'Alemberts – an old nomenclature that is a little out of step with the fuller view of this remarkable complex – are separated by foreland troughs, partly filled with 'lakes' of lunabase, called on the near side after the seasons of the year: Mare

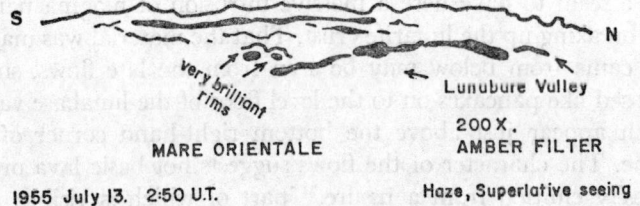

S ———— N

very brilliant rims

Lunubure Vulley

200 X.

MARE ORIENTALE   AMBER FILTER

1955 July 13. 2:50 U.T.   Haze, Superlative seeing

Fig. 13. Telescopic view of Mare Orientale at favourable libration. From an observational drawing by the author.

Hiemis, Vertis, etc. Having been thus christened from views of extreme foreshortening (Fig. 13), these 'maria', too, are not readily identifiable (Plate 5). There is a further lunabase area, clearly visible in the *Zond III* photographs, which overlaps the third mountain ring south-west of Mare Orientale and which the Russians have named the 'Sea of Peace' (*Mirnoye*) in their 'farside' map.

An inspection of the beautiful *L.O. IV* views of the region, and frame HR 187, 2 of 3 in particular, leaves no doubt that the central subsidence of Mare Orientale was accompanied or followed by crustal dilatation farther out. This is present on a smaller scale in the border grabens of Mare Humorum as well, but here it has given rise to conspicuous shifts of crustal blocks, smaller than but comparable to those that have disrupted the Alps and the Caucasus in the development of Mare Imbrium. These shifts, including some rotation, are clearly posterior to the elevation of the Rook and other girdle mountains, which they have parted in the manner of the Alpine Valley. They are also posterior to some of the faults and fractures that cut the mountains, for the lines of both the ridges and the fractures on the two sides of the lunabase valleys fit together like pieces of a jigsaw puzzle.

The land, however, did not come to rest at this stage, and there are other fractures that pass concordantly through the heights and troughs alike, and so must be younger than the latter. The subsurface layers must have remained plastic for some considerable time. Thus the left-hand top part of the frame shows clear signs of crustal folding, which is extremely rare on the Moon, and here bears witness to compressive stresses attendant on the opening of the valleys and their filling with the molten material surging up from the interior.

We seem to have here a massive intrusion of magma parting and breaking up the lunarite crust. That the material was magma and came from below may be seen from the late flows, superimposed like pancakes on to the level floor of the lunabase valley, which appear just above the bottom right-hand corner of the frame. The character of the flows suggests hot basic lava of low viscosity emitted from a fissure,[31] part of which is visible. It is instructive to compare the smooth 'finish' of these flows and of the valley floor with the frozen lake of lava inside a crater a little to the west of the pancake flows. The lake surface is anything but smooth, although nothing like so blocky and cracked as the interior of Tycho (p. 128). The crater itself is clearly a caldera of collapse; its outer glacis are quite steep and high, the interior stands considerably above the level of the surrounding country, and a dimpled parasitic craterlet is perched on the rim. Quite obviously either the composition or the conditions of emission of these materials must have been very different – probably both.

The wide halo of disturbance extends far beyond the triple

collar of mountain ranges. Within this the old craters have been overflowed and partly effaced by vast floods of lava and fluidized ash, but toward the outer edges the character of the ground presents some intriguing features, which are not easy to interpret.

If we take the *L.O. IV* frame HR 173, 2 of 3 the resemblance to Icelandic basalt flows is beyond cavil. The lava has filled in the spaces between the craters and then spilt in over the walls, banking up inside in jumbled mounds. Where exactly it has come from, is not quite clear, but fissure eruptions appear to be the most likely explanation. The whole region, however, has been deeply disturbed, magmas rising along the lines of tectonic weakness and old rings being re-activated as eruptive centres. As in the example considered higher up, many of them cradle lakes of lava, without their walls having been overflowed or breached. The general impression is of flow directed radially away from Mare Orientale, but this may be illusory and due simply to the decrease of activity in this direction. Thus in the middle right part of the frame other directions of flow are readily discernible. Nevertheless, there is no mistaking the seemingly radial striae. What are they?

If we consider the interior of the large ring on the top right of the frame we cannot help noticing that the banks of lava in its eastern (right) half are bounded by ridges concordant with the striations. This cannot be due to directional flow, but only to extrusion from fractures. In other words, the striae can be identified as surface fractures, sometimes manifesting themselves as depressions, into which the overflowing lava has subsided, and sometimes as ridges, where extrusion has taken place.

We shall see later on that this may not be the whole answer; but the surface around Mare Vaporum and beyond bears a somewhat similar striated 'treebark' appearance, which is related to Mare Imbrium in the same way. Here the effect is due neither to a 'liquid splash', caused by impact, nor to ploughing by horizontal ejecta. In fact, no flow or movement is involved at all, and the striae are defined by tholoidal ridges, or *aretes*, as these have been called by O'Keefe.[25] He has made a careful examination of similar features in the *Ranger VII* photographs of Mare Cognitum, where they stand in isolation and bear no directional relationship to Mare Imbrium or any other mare, while their coincidence with the grid trends is clear.

The photographic frames reviewed above cover relatively small portions of the lunar surface, and a global view of the region, such as is presented in the *Lunar Farside Chart,* may serve to correct some misapprehensions. The Mercator projection shows at a glance that the extensive fracturing around Mare Orientale conforms to the main grid trends, which are only occasionally and incidentally subradial to the Mare. In fact, in the west most of the lineaments run parallel to the walls of the outermost mountain rampart, which is much lower on this side (Plate 6).

The response to the stress has varied. In the north-west and west numerous crater chains have developed along both the grid lineaments and the walls of large mountain rings, thus, incidentally, reaffirming the conclusion that these are based on, or associated with, annular fractures and ring dykes (p. 131). On the other sides, however, and in the east and south-east in particular, crater chains are absent and vast amounts of volcanics overly the terrain. Yet, as we approach the borders of the disturbed region on the Earth's side, near Riccioli (HR 173, 1 of 3) and Inghirami (HR 172), the striations assert themselves very strongly, and cannot be fully accounted for by crustal fracturing, which must be expected to decline with the increasing distance from the centre of upheaval. A movement or flow of some kind is clearly involved, because the direction of the striae is affected by vertical relief, which deep-seated fractures are not. In and around Riccioli the lines become tangled and different systems overlap in a way related to the slope of the ground (Plate 18); while in the neighbourhood of Inghirami some of the deposits feather out in the manner of ash flows.

Nevertheless, the change of aspect is gradual and incomplete. North-north-east of the latter ring a few fractures gape open in alignment with the striae, and elsewhere the striae may pass over into indistinct chains of craterlets. Thus two or more different processes appear to have contributed to the total effect, and the connection between them may be largely fortuitous.

The 90-km ring of Inghirami has been invaded from the north-west, but the invasion has stopped short of its centre. Violent moonquakes appear to have shaken the country and re-animated within the ring two intersecting systems of ring dykes, with step-faulting and extrusion. Farther afield, the bowl of Wargentin has been filled to overflowing by an internal flood of lava, triggered, no doubt, by the same disturbance (Plate 13a).

Just north of Inghirami the structure of the ground is ropy, but it would be wrong to equate it with ropy lava, for some of the ropes are over a kilometre thick and are defined not so much by the ridges as by the ditches between them. The 'ropes' wheel over into the undulations of what looks like a conventional lava flow from a 16-km crater some 150 km north-north-west of Inghirami. This smaller crater is undeformed and clearly posterior to the main disturbance. The lava flow bears the appearance of having been 'rubbed over the knuckles', its transverse depressions having been filled in, while its downslope channels have been dredged deeper. Such an effect could have been produced by a flow of water, following that of lava and ash.

The situation near Riccioli is not unsimilar, but the scale is larger. The walled plain, 160 km in diameter, is traversed by a family of graben-rilles sympathetic to the coast of Oceanus Procellarum. Some of them cross the walls, others are confined to the floor. Yet other rilles on the inner glacis may be of the sinuous type, which is difficult to decide, as such flow of water as there may have been will have largely followed tectonic features. Most of the floor is light-coloured, undulating rather than rough, with a few craterlets, mostly dimpled, and at least one dome. It is amply striated in the SW–NE direction, but there is also some faint wavy ridging at right angles to this trend, especially in the south-west. In the northern half of the walled plain, however, spreads an extensive smooth, moderately dark area, with only a few dish-type craterlets. The graben-rilles and striae alike seem to pass under it and disappear, although the large central graben-rille is faintly traceable beneath the dark material, which is obviously of no great thickness.

The distribution of the light-coloured striated material deserves attention. It has obviously arrived from the south-west and crossed the walls of Riccioli. The inward walls upstream of the flow are fairly clear of overlay, with a suggestion of roughening at the top. Beyond them the overlay spreads evenly over the floor, with a few hummocks, banks up slightly against the downstream walls of the minor interior craters, but moves on substantially unimpeded, until the opposite glacis are reached, where it collects athwart in a great jumble of crescentic or S-shaped mounds. The mounds are downstream of the dark flat, and this spreads widest and is presumably founded deepest towards them (Plate 18).

An ash-flow may produce similarly lined deposits, but it would not pile up in mounds in this way, while lava may satisfy the second condition, but not the first.

A careful examination of the dark flat will reveal in it chains of broad and shallow irregular depressions of the type encountered in association with sinuous rilles and interpreted as lake beds. In fact, the whole flat appears to be the bed of an ancient lake, and its dark material to derive from this rather than any outpouring of lunabase lava, which the conditions do not seem to warrant. At this stage we may turn our attention to 'Object A' on the averted hemisphere (p. 151). The peculiarity of this huge ring is the shape of its inner glacis : instead of being laminated or terraced as is usual, they are vertically grooved in a way which vividly recalls the structure of the seaward slopes of the Andes in Tierra del Fuego as photographed from a *Gemini* spacecraft. The downward grooving of the Andean mountain-sides is, of course, the work of running water, and like effects are as a rule due to like causes. In addition, the dark floor of 'Object A' is adorned with a bright cloud-like pattern, suggesting not an emission from a vent or fissure, but deposition from a fluid, such as of silt in a lake (Plate 13b). This suggestion is underlined by the existence of a short but deep sinuous rille spilling into the ring across its north wall (*Orbiter* frame II-33M).

We have had it drummed into us for years that the Moon is and has always been dead; but prejudice has no place in scientific reasoning, where evidence alone counts, and the evidence is strong that both the dark floor of 'Object A' and the Riccioli 'flat' held lakes once upon a time. And, after all, if the sinuous rilles were rivers this is hardly surprising. We may, further, recall that the Moon rotates very slowly, so that severe night frost must be expected even under such an atmospheric cover as it may have enjoyed during the 'water episode'. By the same token the days would be very hot and daytime evaporation intense under a low barometric pressure. In combination these two factors favour night-time precipitation in the form of hoarfrost or snow, depending on the amount of water vapour present in the atmosphere, and if there is enough of it – glaciation. Moreover, as indicated on p. 000, in lunar conditions water and ice must be regarded as volcanic materials, and they may be erupted or extruded from fissures just like lava.

We will remember Tsiolkovsky's curious aprons (p. 85), which suggest 'rock glaciers'. Their striations are closely comparable to those in the Riccioli area. This appearance may be accounted for by a slow and steady movement of some heavy plastic material, or rheid, such as glacier ice. To obviate any possible misunderstanding, let me make it clear that what is meant here by *striae* is not scratches worn out in the rocks by stones and boulders embedded in a glacier, but the grooved and ridged surface of the land. The latter is a large-scale effect, but very similar striation will be seen in the aerial photographs of northern Canada, recently scoured by an ice sheet (Plate 12a). The assumed Orientale glaciation, however, would be combined with intense volcanic activity, in which vast amounts of ash were undoubtedly emitted, so that we must think of glaciers buried under a thick overlay of ash, lapilli and other pyroclastic materials. We must thus expect something like the Tsiolkovsky aprons, and with the underlayer of ice gone, which would be a gradual process, the volcanic coat riding upon it would simply settle down, perpetuating the original striated pattern.

If we reinterpret in these terms the general appearance of the Riccioli region everything falls neatly into place. The glaciers descending from the high ground around Mare Orientale, whose mountain ramparts may exceed 7,000 metres, will have bulldozed down the surface relief, though less effectively than on the Earth owing to reduced gravity, overflowed the walls of Riccioli, tearing away some of the rocks off the top of the inward glacis, spread a carpet of moranic debris over the floor, until the ice has reached the opposite walls, where it will have been halted, wearing out the basin of the future melt-water lake and depositing the mounds of the terminal moraines.

In other words, the blanket of material spreading over the outer margins of the Orientale halo would be basically glacial drift, though almost wholly of volcanic provenance.

The sequence of events would, then, be : tectonic movements (of whatever initiation) originating the basin of Mare Orientale, with tremendous outflows of lava and elevation of the triple mountain girdle; rejuvenation of the surrounding region, progressively declining outwards, lava yielding the pride of place to ash, and ash to gas and water. All this activity is accompanied by an enormous emission of gas, which builds up to a temporary atmosphere of sufficient density to allow liquid water to appear and glaciers to form on the surface. The central portion of the halo becomes wreathed with ice sheets on the Earthward side, while volcanic activity subsides. In the last phase the atmosphere is gradually dissipated, the supply of juvenile waters is sapped, the glaciers melt away and evaporate, leaving behind a blanket of drift.

The existence of sinuous rilles on the mountains and the luna-base 'lakes' around Mare Orientale provides incontestable proof that its formation was followed by a 'water episode', of which there must have been several in the long history of the Moon since its capture by the Earth. Indeed, there are indications of similar developments in the Cognitum-Nubium region, where Fra Mauro and the neighbouring formations are partly buried under thick deposits that look very much like glacial drift, and faint striation is likewise present. This would presumably relate to a much earlier epoch and be associated with the formation of Mare Imbrium, divided by a long interval of time from the birth of Mare Orientale. The sequence of events outlined above, too, may have to be measured in millions of years.

In considering the development of Mare Imbrium, Fielder has reached a similar conclusion.[3] In particular, he finds that the craters in the region of Mare Vaporum have experienced a longitudinal deformation which not only declines away from Mare Imbrium, but also with the age of the craters as judged by the freshness of appearance, the youngest craters being undeformed. This points to a prolonged period of declining stress associated with the formation of the Mare.

In 1962 Brian Warner[35] sought to distinguish two stages in the development of the maria: a compressive and a tensional one, corresponding to the formation of the wrinkle ridges and rilles respectively. Seeing, however, that wrinkle ridges occur typically in the interior of the maria and open fractures on their margins, this distinction does not seem necessary (Firsoff, 1963),[11] as sub-sidence in the central region will generate compressive stress there, in the same way as in a syncline, simultaneously with a tensional stress at the edges (see Fig. 14a, b). Fielder also points out (1965)[5] that the level of the wrinkle ridges is generally below that of the floors of the marginal graben-rilles, so that on the principle of connected vessels the magma expelled from the fractures that have given rise to the former would not rise to the surface in the latter.

None the less, there may be something in Warner's idea, after all, for we have seen on the example of Mare Orientale that the uplift of the mountain girdle, which is a compressive development, was followed by dilatation and the opening of the lunabase rift-valleys which have parted the mountains.

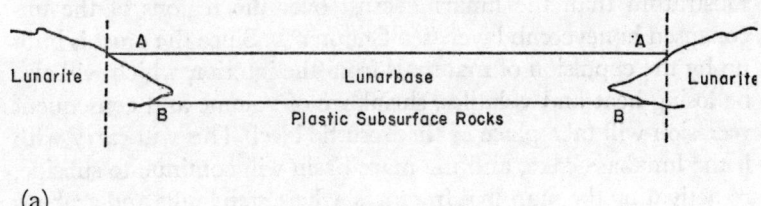

(a)

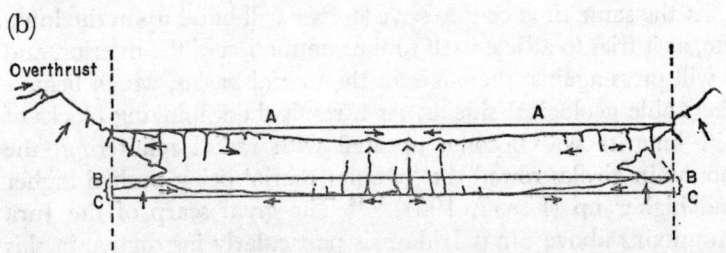

(b)

Fig. 14(a). Represents the situation prior to global shrinkage and crustal shortening. Between A and B a 'cake of lunabase' contained by lunarite 'shores', rests upon relatively plastic subsurface rocks. Lunarite is comparatively light and brittle; lunabase comparatively heavy and tough.

(b). Shows the situation after the shrinkage, in which the base level B sinks to C. The lateral pressure of lunarite on lunabase has comparatively little effect on the latter; lunarite adjusts itself to crustal shortening by up-and-over thrust. The lunabase 'cake' sags in the middle, but is kept up at the 'shores' by upthrusting lunarite. The superficies of the 'cake' increases, with compressive stresses and ridging in the middle and tensional stresses and fracturing on the periphery. At the base level the stresses are reversed and in the middle fluids (magma, water, gases) ascend towards the surface. The direction of stresses is shown by arrows. (See ref. 11).

In the case of Mare Imbrium the great shear movements in the Caucasus, the formation of the Alpine Valley and possibly of Mare Frigoris may correspond to the latter stage in the history of Mare Orientale. The story, however, cannot be read as plainly because of the encroachment by the later neighbour, Mare Serenitatis.

In any event the marial lunabase must be regarded as a heavier and more compact type of rock than the terral lunarite, and so more responsive to subsidence and less to lateral compression. As shown in the diagram, it will be more closely knit to the isostatic substratum than the lunaritic crust over the regions of the un-collapsed honeycomb layer (see Chapter 5). Since the crust is built up by the expulsion of materials from the interior, which will also be losing heat and volatiles, shrinkage of volume and consequent recession will take place at the isostatic level. This will carry with it the lunabase cake, and the mare basin will continue to subside, re-activating the marginal fractures, where step faults and grabens will tend to form and grow. This is, in fact, what we find at the foot of the Apennine scarp in Mare Imbrium and on the borders of Mare Serenitatis.

At the same time compressive stresses will build up in the luna-rite, as it tries to adjust itself to the contraction of the interior, and it will press against the edges of the marial basins, which behave like stable geological shields. As a result the impinging blocks of stiff lunarite will become uptilted with radial splintering, the mountain girdles round the 'circular maria' being pushed higher and higher up (Firsoff, 1961).[9, 10] The great scarp of the Jura Mountains above Sinus Iridum is particularly instructive in this context. It is the rim of a comparatively undifferentiated crustal block, tilted away towards Mare Frigoris, and rises to at least 6,000 metres at an average angle of at least 30° over the floor of the Sinus. If we regard the latter as an erstwhile walled plain partly engulfed by Mare Imbrium (a somewhat doubtful inter-pretation), then the height of the scarp, its structure and steepness are altogether too great for a ring formation of these dimensions (diameter, 250 km). Such a feature could arise only by continued uplift.

It has already been pointed out (p. 72) that the Jura Moun-tains differ from the Alps, Caucasus and Apennines in not having an upraised edge, while the high peaks of the latter ranges are of volcanic origin and stand considerably above the hinterland block. We thus have here a combination of tectonic uplift with mag-matic action along the border fracture or fractures. This has an exact parallel on the Earth in the Andes, similarly elevated and kept volcanically active by the steady drift of the South American continent against the unyielding 'shield' of the Pacific Ocean.[18] This is one more point of resemblance between our oceans and the lunar maria, whose wrinkle ridges are structurally similar to the basic oceanic ridges, such as the Mid-Atlantic Ridge.[5] The irregular maria are also regions of crustal collapse, but the absence of underlying mascons serves to confirm the impression that the disturbance is less deep-seated, whether we do or do not accept von Bülow's description of such 'epicontinental maria' as lunabase-flooded terra.[1] To some extent this is certainly true, as, for instance, in Mare Nubium submerged lunarite formations can be traced with comparative ease. Nevertheless, I think that the analogy with the Thulean basalts or Deccan traps can be pushed too hard, and the alternative proposed in *Surface of the Moon*,[9] whereby we have here not so much an effusion of lava or ash as a widespread metamorphism, or pneumatolysis, by hot fluids permeating the crust from the interior, deserves serious consideration. Be this as it may, the general absence of mountain ranges along the shores of such maria, on the one hand, and the profusion of wrinkle ridges within them, on the other, may be taken as an indication that the cake of the lunabase is too thin to offer effective resistance to lateral compression. The difference, however, is one of degree.

The catastrophic period of capture apart, the smaller size of the lunar globe should slow down all crustal movements, with long periods of quiescence and growing stress culminating in shorter epochs of activity and adjustment that correspond to our major orogenies. Still, even on the Earth an average interval between these is 250 million years.

At the moment the Moon appears to be passing through a period of quiescence, but this does not mean that all selenological activity has ceased. There is nothing intrinsically improbable in the reported changes in the appearance of the small crater Linné in Mare Serenitatis, but the verdict has to be 'not proven' for the time being.[22] On the other hand, N. A. Kozyrev[21] not only observed visually, but obtained a spectrographic record of, a gas eruption in Alphonsus at the Crimean Astrophysical Observatory in November 1958, and this cannot be dismissed. There have been other observations of similar 'lunar transient phenomena', which we shall consider under another heading. In 1968, however, during an infrared survey of the Moon, Hunt, Salisbury and Vincent[19] discovered a thermal anomaly in the faulted region of Mare Humorum, between Gassendi A and Liebig F. Here the surface 'is warmer than its surroundings during the lunar afternoon, which is the sort of thermal behaviour one would expect of an internal heat source'. The marginal fractures of Mare Humorum would thus be magmatically active, and a concomitant movement along the fault-lines may be suspected.

In 1964 O'Keefe[25] drew attention to a group of dark mounds in the *Ranger-7* frame 176. The mounds are aligned with some other similar, light-coloured tholoids in the vicinity, but, unlike these, are unaffected by the material from the rays of Tycho and Copernicus, and so posterior to these. Since now the rays are believed to be among the youngest selenological features, the igneous activity in Mare Cognitum which has raised these dark mounds must be quite recent.

Much of the lunar surface is thinly coated with unconsolidated, or lightly coherent, material, already mentioned on p. 127, resting upon harder formations. The situation is well illustrated by the rolling tracks of boulders on the slopes of Vitello, photographed by the *Orbiter V*. The larger boulder has bumped from side to side on striking the harder substratum. But similar tracks are to be seen also in the Hyginus Rille and on the sides of Schroeter's Valley.

The boulders cluster round the top edges of smaller craters and rilles, but are relatively scarce on their floors, where the erosive agencies producing boulders do not seem to operate. In particular the floor of Schroeter's Valley must have originally been completely clear of rock debris, which may have been swept away by running water, so that any boulders found there have rolled down from the sides. The interesting point is that very few of these have left visible tracks (cf. *L.O. V* frame HR 204, 1 of 3). A few rocks are just as clear as on Vitello, while some others can be made out with difficulty. Obviously something is at work, to remove these tracks. This may be meteoric churning, but Fielder[4] calculates the rate of erosion by small meteoroids at between 4×10^{-11} and 10^{-6} cm per cm^2 of the surface per year, which seems to be far too slow to produce the desired effect. On the other hand, the whole area is covered with ash from Aristarchus, and continued showers of this fit the facts better.

If the fine material were of cosmic provenance it would average the same colour over the whole of the Moon, in Rudaux and de Vaucouleurs' simile 'like dust in a long-neglected household'. This is far from the case. Not only is lunabase much darker than lunarite, but sharply delimited areas of different shade exist within it. The material must, therefore, be local. True, it was proposed by Thomas Gold in 1955[16] that the Moon's surface is subjected to erosion by corpuscular and short-wave radiation of the Sun, resulting in very fine dust. Glass, which is a silicate, darkens to a violet shade on prolonged exposure to short ultraviolet rays, and many other rock materials share this property of *tenebrescence*. Thus, argued Gold, this dust will be dark. Gravity, meteoric impacts, electrostatic repulsion and Brownian movement (this requires the presence of gas, but some of the dust is assumed to be so fine as to behave like a very heavy kind of gas), will gradually remove this dark dust from the exposed heights into hollows, and the marial basins, being the deepest of all these, will be the natural areas of catchment for it, whence their dark colour.

Theoretically some such effect should be present, and the idea was taken seriously, for instance, by Brian Warner (1961).[33, 34] But observational objections to it are many and insuperable (Firsoff, 1959),[8] and in the present state of lunar knowledge it can no longer be entertained. To take one example, what is supposed to have become of all the dark dust generated over the long selenological ages on the far side of the Moon?

The usual alternative for lunabase are lava, ignimbrite and mud,[7, 32] the two latter whereof seem to have been first suggested by myself (1959),[7] while in Gilvarry's opinion[15] the maria were once real seas, so that their present surface would also be mud or perhaps sand. We have seen that the 'lunabase' flat in Riccioli may be mud, although the material of lunabase valleys around Mare Orientale is quite certainly lava. The wrinkle ridges are dark, and so, incidentally, are O'Keefe's[25] mounds in Mare Cognitum; both are definitely lunabase extrusions. That the extrusion was liquid or plastic can be clearly seen in, say, the *Orbiter III* views of Oceanus Procellarum in the region of Marius. The association of wrinkle ridges with elementary rings (Guest and Fielder, 1968)[17] corroborates their extrusive character. Since the composition of lunabase seems to approximate that of our basalts (p. 51), O'Keefe's[27] argument against basic lava being able to form 'cumulodomes' falls to the ground. It would, in fact, seem that those flows of lunabase that have been emitted under atmospheric cover behaved like our basaltic lavas and spread fast, far and thin; but in airless conditions they frothed up and caked over before making much headway. This would explain the smoothness of the maria and the interiors of such rings as Gassendi, Plato and Tsiolkovsky, on the one hand, and the roughness of the lunabase floors of some other, usually smaller, craters.

However, the three alternatives are not necessarily mutually exclusive, and O'Keefe and Cameron[24] argue most persuasively in favour of ash-flows, setting to ignimbite. According to their calculations in lunar conditions fluidized ash would continue to flow much longer and progress much farther than on the Earth. Thus a terrestrial flow of 80 km would correspond on the Moon to one of 2,700 km, and the duration of 0·8 hours to 27 hours. It may be unwise to pin too much faith on these figures, but they have a sound basis. The only difficulty is the basic composition of lunabase. Basic ash-flows are extremely rare on the Earth, only one such having, in fact, been discovered in a survey carried out by Ross and Smith.[28] On the other hand, low gravity and barometric pressure would encourage the fragmentation of gas-rich lavas, and basic lavas contain a high percentage of water, which would behave as gas, so that this objection may not carry much weight.

One argument advanced by O'Keefe in support of the ash interpretation is the subdued appearance of some craterlets alongside of others which are cut sharp and clear in the close-up photographs of the surface (p. 67). Lava overflowing pitted ground would fill in all its hollows and set to a dead level, but ash, although laid down smooth to begin with, is differentially compacted where thickest, so that the old bumps and dips reappear after a time in subdued relief.

We have, of course, man-height views of marial ground from the soft-landed probes. The *Luna IX* landed in Oceanus Procellarum, 70 km north-east of the mountain ring Cavalerius, in a slightly upraised brightish area. The photographs taken by the probe show a rough, harshly-porous surface on the centimetre scale, agreeing with Jan van Diggelen's finding that the lunar reflection curves are best matched by those of the false reindeer moss (*Cladonia rangiferina*).[30] But in the *Surveyor* close-ups both from Oceanus Procellarum (Flamsteed P) and Mare Tranquillitatis the texture of the ground, while not unsimilar, is much softer, and in the first instance there is definitely an overlay of lightly consolidated dusty matter a few millimetres thick. Winifred S. Cameron[2] suggests that the difference may be due solely to the angle of illumination, but I think there is more to it, for the Jodrell Bank version of the first *Luna IX* picture, which is sharper than the Russian, clearly shows thin linear fractures of the ground (Plate 19a).[12] It must thus be brittle and free from any soft overlay, probably an unmodified surface of highly vesiculated lava.

The impact of the *Surveyor I* landing pads and the digging operations carried out by the *Surveyor III* indicate that the surface has the mechanical strength of wet sand and consists of loosely cemented grains of the order of 1 millimetre. This is compatible with a flow of ash, but could also be produced in lava by meteoric churning, sputtering by solar and cosmic corpuscular radiation,[4, 34] molecular creep, and/or the erosive agency of alternate sorption and desorption of gas (Firsoff, 1959),[7] which is seldom taken into account and which we shall reconsider in another context.

This is a problem that the expected lunar landing should be able to solve, although we have just seen that the surface of the Moon may differ from place to place, even in the maria.

On the basis of the rate of micrometeoric erosion quoted earlier on (p. 172) and the comparable surface wear due to cosmic rays, Fielder concludes that 'no lunar mare is older than 6×10^8 years'.[4] Numerical results provide a welcome check on qualitative thinking, and mathematics is a wonderful tool of the mind. But it is not a magic, and for the tool to be effective we must be sure of all the relevant facts and relationships, which is not quite the case here. The besetting sin of all such calculations is that they do not allow for any possible change in conditions. In this particular case it seems that the history of the maria involves several periods during which the Moon possessed an appreciable atmosphere, which spreading high up under low gravity would form a most effective bumper against the meteoric bombardment and cosmic rays alike. On the other hand, some subaerial denudation must likewise be taken into account. The problem cannot be solved as simply as that. Still the assumed rates of erosion are probably not too far off the mark. It all depends on what is meant by the age of a mare;

whether it is reckoned from its tectonic initiation or the formation of the present surface. If the first case, most maria, and Nectaris in particular, are probably much older than 600 million years, although Mare Orientale may be a good deal younger.

REFERENCES

1 B ü l o w, K u r d v o n (1964). 'Beiträge zur Selenologie, IV', *Geologie,* **12** (6/7), p. 899 ff.

2 C a m e r o n, W i n i f r e d S. (1968). Private communication.

3 F i e l d e r, G i l b e r t (1962). 'Origin of Mare Imbrium', *Nature,* **193** (4812), p. 258.

4 F i e l d e r, G i l b e r t (1963). 'Erosion and Deposition on the Moon', *Planet. Space Sci.,* **11**, p. 1335 ff.

5 F i e l d e r, G i l b e r t (1965). *Lunar Geology.* Lutterworth Press, London.

6 F i r s o f f, V. A. (1957). 'Palaeography of the Nubian and Imbrian Plains', *J.B.A.A.,* **67** (4), p. 130 ff.

7 F i r s o f f, V. A. (1959). *Strange World of the Moon.* Hutchinson, London.

8 F i r s o f f, V. A. (1959). 'Is the Moon Covered with Dust?', *J.B.A.A.,* **69** (4), p. 153 ff.

9 F i r s o f f, V. A. (1961). *Surface of the Moon.* Hutchinson, London.

10 F i r s o f f, V. A. (1963). 'Selenological Implications of the South Australian Ring Structures', *Nature,* **198** (4875), p. 78 f.

11 F i r s o f f, V. A. (1963). 'The "Compressive" and "Tensional" Stage in Lunar Maria', *J.B.A.A.,* **73** (3), p. 114 ff.

12 F i r s o f f, V. A. (1967). 'Close-up Views of the Lunar Ground', *J.B.A.A.,* **77** (4), p. 251 ff.

13 F i r s o f f, V. A. (1968). *Exploring the Planets.* Barnes, Cranbury, N.J.

14 F i r s o f f, V. A. (1968). *New Scientist,* **37** (587), p. 527 ff.

15 G i l v a r r y, J. J. (1960). 'Origin and Nature of Lunar Surface Features', *Nature,* **188** (4754), p. 886 ff.

16 G o l d, T h o m a s (1955). 'Lunar Surface', *M.N.R.A.S.,* **115** (6).

17 G u e s t, J. E., and F i e l d e r, G. (1968). 'Lunar Ring Structures and the Nature of the Maria', *Planet. Space Sci.,* **16** p. 665 ff.

18 H o l m e s, A r t h u r (1965). *Principles of Physical Geology.* Nelson, London and Edinburgh.

19 H u n t, G. R., S a l i s b u r y, J. W., and V i n c e n t, R. K. (1968). 'Infrared Images of the Eclipsed Moon', *Sky and Telescope,* **36** (4), p. 223 ff.

20 K o p a l, Z d e n e k (1966). *An Introduction to the Study of the Moon.* Reidel, Dordrecht.

21 K o z y r e v, N. A. (1959). *Sky and Telescope,* **18** (4), p. 42 ff.

22 M o o r e, P., and C a t t e r m o l e, P. (1967). *The Craters of the Moon.* Lutterworth Press, London.

23 MULLER, M., and SJOGREN, W. L. (1968). *Sky and Telescope,* **36** (4), p. 219.

24 O'KEEFE, J. A., and CAMERON, W. S. (1962). 'Evidence from the Moon's Surface Features for the Production of Lunar Granites', *Icarus,* **1** (3).

25 O'KEEFE, J. A. (1964). 'Interpretation of Ranger Photographs', *Science,* **146** (3643), p. 514 f.

26 O'KEEFE, J. A. (1966). 'Lunar Ash Flows' in *The Nature of the Lunar Surface. Proceedings of the 1965 IAU–NASA Symposium,* p. 259 ff. Johns Hopkins Press, Baltimore, Md.

27 O'KEEFE, J. A., LOWMAN, O. D., JR., and CAMERON, W. S. (1967). *Science,* **155** (3758), p. 77 ff.

28 ROSS, C. S., and SMITH, R. L. (1961). *Ash-Flow Tuffs: Their Origin, Geologic Relations and Identification.* U.S. Geological Survey, Professional Paper 366.

29 SPURR, J. E. (1948). *Geology Applied to Selenology – III – Lunar Catastrophic Theory.* Rumford Press.

30 STRUVE, OTTO (1960). 'Photometry of the Moon', *Sky and Telescope,* **20** (2), p. 70 ff.

31 TYRRELL, G. W. (1931). *Volcanoes.* Butterworth, London.

32 UREY, H. C. (1967). 'Study of the Ranger Pictures of the Moon' in 'A Discussion on the Physics of the Moon and Its Environment', *Proc. R.S., Series A,* **296** (1446).

33 WARNER, BRIAN (1961). 'The Lunar Maria', *Planet. Space Sci.,* **5**, p. 283 ff.

34 WARNER, BRIAN (1961). 'Accretion and Deposition on the Surface of the Moon', *Planet. Space Sci.* **5**, p. 321 ff.

35 WARNER, BRIAN (1962). 'Some Features of the Lunar Grid System', *J.B.A.A.,* **72** (4), p. 181 ff.

Abbreviations used

J.B.A.A.	= Journal of the British Astronomical Association.
M.N.R.A.S.	= Monthly Notices of the Royal Astronomical Society.
Planet. Space Sci.	= Planetary and Space Science.
Proc. R.S.	= Proceedings of the Royal Society.

13

Seeing is believing

Not so long ago any report of a so-called 'lunar transient phenomenon' involving a temporary change in the visibility, appearance, brightness or colouring of a local surface feature, if not dismissed out of hand, would have been regarded with extreme suspicion. But fashions in thought are almost as fickle as fashions in clothes, and today the LTPs have suddenly become respectable, much observed and recorded. One is even inclined to wonder if the pendulum has not swung too far in the opposite direction.

This change in attitude is clearly traceable to Kozyrev's observations of red glows, accompanied by a distinctive line spectrum in emission, near the centre of Alphonsus on 3rd November 1958 and again on the 23rd October 1959.[14, 15] The operative factor was the spectrographic record, which alone could objectively establish the occurrence of a 'lunar event'. Yet, as often happens, there are earlier observations of lunar spectra. Thus an 'unusual spectrum with strong absorption in yellow' was seen by Airy at $23^h 10^m$ during the lunar eclipse on the night of 23rd – 24th August 1878. In the following year several observers at Melbourne, Australia, recorded changes in the solar spectrum at the Moon's limb, suggesting absorption by a lunar atmosphere, during a solar eclipse at $8^h 16^m$ on 2nd February.[19] Admittedly, these observations would seem to relate to the whole Moon rather than any local event, and in the first case the source of absorption could have lain in our own air. Still, the case is at best 'not proven' and seems well worth investigating. I will return to this in a further context.

Kozyrev's observations did not stand alone. Thus Poppendick and Bond saw a 'diffuse cloud over the central mountain' of Alphonsus on 19th November 1958, and on the same night Wilkins and Hole noticed a red patch there. This, being about a fortnight after Kozyrev's 'gas eruption', suggests continued activity. Significantly a red patch was also seen by Warner on 6th January 1960, following Kozyrev's second observation.[17, 19, 20]

Be this as it may, 579 reports of LTPs, dated from 1540 to 1967, have been distilled, with some rejections, from various sources, including a modest contribution by myself, and catalogued by Barbara M. Middlehurst *et al.*[19] It is not always easy to weigh such reports, but even if we reject 50 p.c. of the listed observations as doubtful – and this seems going a little too far – there is more than enough left for serious consideration.

The two earliest items antedate the invention of the telescope and are correspondingly startling. At five in the morning of 26th November 1540 some burghers of Worms saw a star shining on the dark side of the Moon, and an anonymous English entry, tentatively dated 5th March 1587, also speaks of 'a sterre . . . sene in the bodie of the mone . . . whereat many men merueiled. . .' The star appeared between the cusps as in the Muslim symbol, which, too, has been regarded by some as an indication of a similar occurrence in an unrecorded past. Another report of a star on the dark side is post-telescopic and comes from 'several New Englanders' under the date of 26th November 1668. On 3rd May 1715 no less a man than Edmund Halley in England, and independently de Louville in France, saw 'lightning' on the Moon during an eclipse, which the latter explained as 'storms'.[19] This reads somewhat like a compilation by Charles Fort, but, after all, Fort did no more than collect records of unusual events, which may or may not be true.

We have several eighteenth-century and later reports of 'active volcanoes' on the Moon. Thus Sir William Herschel saw in March 1783 bright spots alone and on 4th May of the same year with Mrs. Lind a red spot, like a 4th-magnitude star and less than 3″ in diameter, near the crater Aristarchus. In 1787 Herschel again observed three bright spots on the dark side of the Moon in March, and on 19th April what he describes as 'three volcanoes'.

The brightest of these, located 3' 57" from the north limb, appeared to him like 'a small piece of burning charcoal covered by a very thin coat of white ashes seen in faint daylight'. The other two were nearer the centre of the disc. Next night they were still there, the brightest 'volcano' in increased brilliance. Schroeter saw a bright spot on the night side in 1788. Bright spots, sometimes of a peculiar fiery colour, have been seen in the area of Plato: in the Teneriffe Mountains and close to Mt. Blanc in the Alps, at various times, spread over more than a century. On three nights of 4 – 6th May 1821 Ward and Baily observed a bright spot on the night side in the vicinity of Aristarchus.

Here a little caution is indicated, as Aristarchus can often be made out quite clearly in Earthshine on the dark side of the Moon. I also remember seeing it like a star through a cloud that completely obliterated the rest of the Moon. Or was this an LTP? In telescopic observation a ghost image of the Moon, and during an eclipse – of the Sun, or other reflections from the lenses may become projected on to the dark side of the Moon and create a stellar appearance. This may mislead an inexperienced observer. Such an accusation, however, can hardly be laid at the door of Herschel, who was very precise about locating his 'spots', and most of the other names on the list command comparable respect. On the other hand, there may be a good reason why bright spots on the dark side are seldom reported in recent times. With much improved instruments and higher powers only a small portion of the Moon's disk is observed at a time, and few people bother to look at its unilluminated part. This presents a serious disadvantage in hunting for 'transient phenomena', where lower magnifications and higher surface brightness are preferable, especially when monochromatic filters which cut off most of the light are employed.

We have, however, strong recent corroboration of activity in Aristarchus and Schroeter's Valley. On 30th October 1963 J. A. Greenacre and E. Barr, who were engaged at the Lowell Observatory on mapping work, noticed reddish-orange glows declining to ruby-red in the Cobra-Head and along the rim of the crater. The glows endured for 25 minutes. On 28th November the glows reappeared, and on this occasion were also seen by John Hall, Director of the Observatory. The glows turned to violet and were succeeded by a blue haze. The estimated duration of the effect as from the beginning of the observation was 1 hour and 15 minutes. The attempt at obtaining a photograph of the changes proved unsuccessful, but these were confirmed independently by P. Boyce, observing with the 72-inch reflector of the Perkins Observatory, and similar appearances on the same date were reported by C. Tombaugh and W. Fisher as well.[18, 20]

It is these observations that have put the final seal of approval on the LTPs.

A blue, or blue-violet, haze has often been seen in or near Aristarchus by various observers, notably Bartlett. When the Moon is full and high and the air clear the diamond-shaped area north-west of Aristarchus, including Schroeter's Valley, is clearly tinged with a mustard yellow.[7, 9] Yet, when I viewed it through a violet filter, which cuts off yellow light, it continued, contrary to expectation, to look bright. A. G. Smith,[25] who observed it with an electronic image-converter, also found it pale in the ultraviolet, which contrasts sharply with the ultraviolet exposures taken in 1926 at Lick by W. H. Wright and Dorothy Applegate,[26] but agrees with my observation. The effect may be accounted for by a thin mist or vapour.

'Seeing is believing', but sometimes we see so uncertainly – or perhaps so unexpectedly – that we tend to disbelieve our eyes, and many astronomers who tend to be most vocal in rejecting or questioning such observations all-too-often know the Moon only from photographs. It is, of course, intrinsically unlikely that a photograph would catch an evanescent effect of the type described above, and even if it did this may not show at all or be very difficult to recognize for what it was. Nevertheless, some of the *L.O.V* close-ups of Aristarchus, and frame HR 195 – 3 of 3 in particular, suggest mistiness. The early lunar observers had the advantage of blessed ignorance and could record what they saw without inhibitions, but we moderns have been so indoctrinated that we find it difficult to accept the reality of a lunar event. The joke, however, may be on us, not on them, for there is nothing intrinsically improbable about a lunar volcano being in action now. There is abundant evidence of lunar vulcanism, as we have seen in the preceding chapter, and the conviction that it has all ceased is a matter of dogma, not fact.

Speaking for myself, I have observed a number of colour enhancements or brightness fluctuations on the Moon, and, since other observers have reported similar effects, I am inclined to believe in the reality of my own observations; but this belief does not amount to absolute certainty. To this general statement, however, there is one exception, not listed in *Chronological Catalog of Reported Lunar Events*.[19]

In July 1955 I followed the appearance of the Aristarchus area from lunar sunrise to sunset with a 6½-inch reflector and a power of 200 diameters, and made a number of drawings. To some extent the variable 'seeing' may affect the result, but my drawings

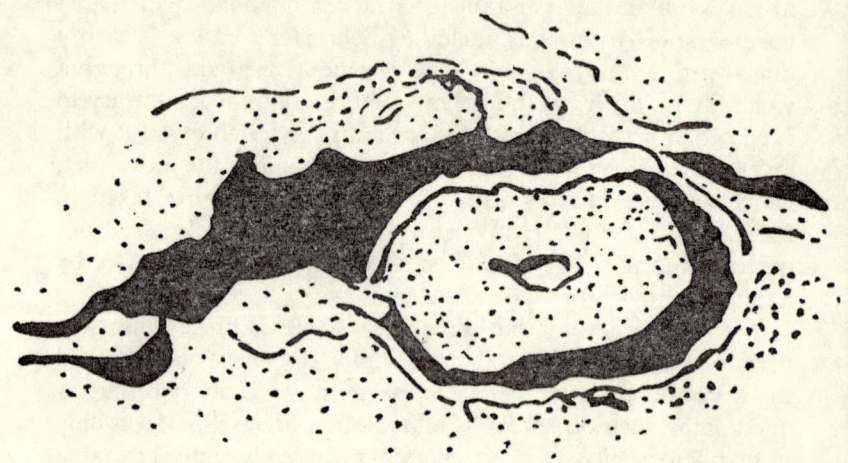

Fig. 15.   A 'pseudo-peak' in Herodotus, seen by the author on 15th July 1955.

are accurate, and that of 15th July, 3ʰ 50ᵐ U.T., shows a central peak casting a shadow (Fig. 15) in Herodotus. Such a peak has been seen by many other observers, and it was there as plain as a pikestaff. I did not regard this as anything remarkable. Yet the *Orbiter* photographs leave no doubt that there is no such thing in Herodotus. Equally can there be no doubt that on that night there was something there, as it were, 'impersonating a hill', and this means an eruption. Another peculiar thing about Herodotus is that quite a few observers 'saw' Schroeter's Valley extending into the crater,[23] which it quite certainly is not. Was this another cloud? This is not unlikely, for W. H. Pickering,[22] observing at the Southern Station of the Harvard University, in

Arequipa, Peru, in 1893, was convinced that he saw on several successive nights a 'vapour column' rising from the Cobra-Head, and F. H. Thornton, using an 18-inch reflector on 10th February 1949, recorded 'a puff of whitish vapour obscuring details for some miles in the area.[9, 19, 20]

The *Catalog* does not list any observations of the central peak in Herodotus, but it mentions a 'temporary hill', 3 km in diameter and casting a shadow, seen by Cragg on 16th July 1964 south-east of Ross D. To be classed in the same category is the report by several observers, published in *Scientific American* in 1882, of 'two pyramidal luminous protuberances' on the Moon's limb. These protuberances were a little darker – presumably less substantial – than the rest 'of the Moon's face' and slowly faded away.

Other localities where clouds or obscurations have been frequently reported are the floors of Plato and Schickard, Gassendi and Mare Crisium. We may instance here Patrick Moore's observation of 2nd August 1939, when the internal detail of Schickard was obliterated by an extensive mist. On 31st August 1944 the floor of the great walled plain looked misty to H. P. Wilkins and some minor craters in it, which are normally well shadowed, stood out as white spots under a low Sun. Of the earlier observers A. S. Williams saw the floor of Plato at sunrise 'glowing with a curious milky kind of light' on 27th March 1882; and E. E. Barnard at Lick found the bowl of Thales filled with luminous haze one night in 1892. Cattermole's report of the disappearance of the central mountains of Eratosthenes on 11th May 1954, although the surrounding detail remained clearly visible, is a comparable phenomenon. I myself[9] saw a faint mist in Theophilus on 25th June 1955; it was gone next night. One of the most peculiar effects I have seen was on 8th September 1954, at about 20^h U.T., when Proclus, viewed through a blue Dufay tricolour separation filter, appeared to brighten up conspicuously, sometimes to dazzling brilliance, for periods from a few seconds to some minutes at a time and again return to normal. Two hours later it no longer showed any fluctuation in brightness through blue or any other filter. It looked as though great waves of gas, laden with crystals or very fine dust, were welling up from the crater. This may, however, be a case of 'lunar glow', possibly involving a different physical principle, a systematic look-out for which has been kept since 1964 under the British and American Moon-blink programmes, sponsored by the NASA.[2, 3, 18]

The method consists essentially in a quick comparison of the Moon's image as seen through an appropriate red and blue filter respectively. The filters are mounted in a rotary disk and can be flicked back and forth. If any surface feature is brighter in either the blue or red light it will 'stick out' of the background or 'blink' when the filter carrier is rotated. Equalizing neutral filters may be used to reduce both images to the same intensity, and in the American, though not the British, version the image is passed through an electronic image converter and appears on a screen in grey monochrome. Colour enhancements of only 2 p.c above the background should be detectable.[3]

I used a substantially identical device since 1952, patented it as from 1954,[6] and carried out a systematic series of lunar observations over a period of three years ending in 1956.[5, 7, 9] But it was not 'news' then! My original instrument was very modest (*see* Chapter 1), and, cutting my coat according to the cloth, I concentrated mainly on the detection of faint colorations of extensive areas of the surface and their relation to the phase, using primarily a set of three filters: red, green and blue-violet (Dufay), which separated the three basic colour responses of the human eye and gave by subtraction images corresponding to the complementary, or printer's, primaries: peacock blue, magenta and yellow. The human eye is an extremely delicate and adjustable instrument of observation, and simple equipment presents various advantages over sophisticated apparatus, so that very interesting results can be obtained by such means. In this case these relate to lunar colourings rather than LTPs, but I did observe a number of red and blue brightenings as a by-product of the main investigation.

It happened not infrequently that the relative brightness of the same group of craters, especially of the rayed type, varied from night to night. The method of detection was the same as in observing variable stars: if, say, crater A was brighter through a red filter than B on a certain night, while both had the same surface brightness on other occasions and C remained unaffected, an LTP red glow must have occurred in A. I did, in fact, detect such a red glow in Aristarchus at 22^h 00^m U.T. on 11th August 1954, and in Timocharis on 8th and 10th October 1954; but on 13th July 1955 Aristarchus showed a blue 'brightening', as did Timocharis on 3rd

August 1955, when it also looked large and diffuse through a blue filter. I made some other similar observations, which I did not think worth reporting for the *Catalog*, but have subsequently made available to Winifred S. Cameron.

To turn to the present-day Moon-Blink programme, I will quote the following account from Patrick Moore, who is actively involved in it (p. 85, *The Craters of the Moon*):[21]

... At 21.30 U.T. on 30th April 1966, Sartory ... detected a strong 'blink' on the south-east rim of Gassendi ... he at once telephoned other observers, and at 22.18 a red glow, in precisely the position of Sartory's blink, was seen by Moore and Moseley, using the 10 in. refractor at Armagh Observatory. The observations were truly independent, since Sartory, again according to plan, had not given the precise location of his 'blink' in order to eliminate unconscious prejudice. On 30th April and again on 1st May, further colour phenomena in Gassendi were seen independently by Sartory at Farnham with the Moon-Blink, and visually by Ringsdore at Ewell and by Moore and Moseley at Armagh; the main feature was a wedge-shaped reddish-orange streak extending from the west wall across to the central peak. The positions and times agree so well that there seems no doubt of the genuineness of the phenomena ...

The joint Moon-Blink programmes have produced a number of similar reports, which cannot be examined in detail in the present context. Glows, however, were seen on the Moon long before any of these systematic observations began. To give a few examples, on 19th February 1885 Gray saw 'a small crater near Hercules' glow dull red 'with vivid contrast' and Knopp observed red patches in Cassini on 21st February of the same year; on 19th May 1912 Valier noticed a small red glowing area on the dark side; 15th June 1913, a 'small reddish spot' seen by Maw in South; 22nd February 1931, a reddish glow in Aristarchus seen by Joulia, and a bluish 'glare' by Goodacre and Molesworth in the same year. It will be appreciated that these observers did not use filters, so that most of the LTPs reported in the Moon-Blink work (extending this term to my own investigations of 1953–56) would probably have escaped notice.

One further observation deserves special mention, not only by reason of the discussion it provoked, but because it is backed up by a photographic record, a red-filter exposure showing an enhanced brightness, which is absent from the photograph taken immediately afterwards in green light. I am referring to the extended red glow seen round Kepler by Z. Kopal and T. Rackham[12] at the Pic-du-Midi Observatory on the night of 1st – 2nd November 1963. This effect, attributed by them to fluorescence, did not recur during the following lunation when a careful search was made for it by G. L. Roberts.[21]

Against all this must be set the wholly negative result of 3,000 hours of 'lunar surveillance' between 1965 and 1968 with the 24-inch reflector of the Corralitos Observatory, in New Mexico, under the direction of J. Allen Hynek. A rotary filter carrier was used mounting four filters with a peak transmission at 3,900 Å (violet), 5,500 Å (green), 7,000 Å (red) and one covering a band of 8,500 – 10,400 Å in the infrared for thermal effects, together with an orthicon tube, giving a monochromatic picture on a screen. Several LTPs were reported during the period of observation in the Moon-Blink programme, so that this result is difficult to understand, unless some undiscovered instrumental difficulty is involved.

On 21st and 22nd April 1967, however, a general brightening of the Moon, strongest at the subsolar point, in blue was recorded at Corralitos. In my own filter work I have noticed considerable variability in the strength of the blue image of the Moon, as also that of Mars. In the latter case this is, no doubt, related to the fluctuations in the 'violet layer', 'blue clouds' and 'blue clearings', and it is not beyond the realm of possibility that similar factors may operate on the Moon. At the time I ascribed the effect simply to the changing transparency of our air to the light of shorter wavelengths, but this could not account for the increase in brightness towards the subsolar point, which would point to a connection with the Sun.

Indeed, the brightness of the Moon at the same phase has been found to vary from month to month and year to year (G. Rougier, 1933). Librations, polar phase and a variable contribution from the Earthshine[11] may be involved to some extent, but there appears to be more to it, and a correlation with the sunspot cycle has been claimed. In any event the photometric studies by Gehrels and associates seem to show that in the period of maximum solar activity from 1956 to 1959 the Moon's surface averaged 10 to 20 p.c. brighter than between November 1963 and January 1964, about the sunspot minimum.[13] The Moon *luminesces*, or emits a certain, variable amount of light in addition to the reflected light of the Sun and the Earth.

This problem first came to the fore in connection with total lunar eclipses, which vary greatly in darkness, that of 1884 being exceptionally dark. *English Mechanic* printed in 1918 an interesting report by several observers that during the total eclipse of 4th – 5th July 1917 the brightness of the lunar disk increased towards the limb. This prompted M. A. Nodon, of Bordeaux, to suggest that the Moon 'may possess a luminosity of its own in the nature of phosphorescence', the enhancement towards the limb being due to perspective.[4] This, it will be noticed, is the precise opposite of the Corralitos observation. The subsequent investigations by Link (1932, 1950) and Cimino (1957) showed that the variable refraction in the Earth's atmosphere was insufficient to account for the observed effects, about 10 p.c. of the light of a full Moon being due to luminescence. There also seemed to be a good correlation with the solar cycle.

Between 1956 and 1959 Kozyrev in the U.S.S.R. and Dubois in France found evidence for a widespread fluorescence of the surface in different parts of the spectrum. Dubois compared the depth of the Fraunhofer lines in the spectra of the Moon and of the scattered sunlight, and according to him most of the Moon luminesces strongly in yellow and red, but the edges of the maria show a distinct green luminescence, while some marial regions also emit blue light. He attempted to associate these effects with certain minerals, known to fluoresce in the corresponding colours, but such identifications do not look very convincing.[4, 16]

Kozyrev studied the profiles of the Fraunhofer lines and the *H* and *K* lines of ionized calcium in particular with the 50-inch reflector of the Crimean Astrophysical Observatory and a dispersion of 50 Å per mm, and found distinct fluorescence with spectral maxima near 3,900 and 4,300 Å (a blue-violet effect) in Aristarchus, reaching a peak intensity shortly after full. The observed emission was far too strong to have been caused by the normal short-wave radiation of the Sun, but he thought that a stream of high-energy protons emitted by a solar flare could account for it.

According to Kopal the spectroscopic work carried out in 1961–62 has generally confirmed the earlier results, establishing the reality of lunar luminescence 'beyond reasonable doubt'.[13]

The same, however, cannot be said about its causes.

A distinction must be drawn between general luminescence, defined non-commitally as an emission of secondary light, and the more intense local glows. Laboratory tests made with terrestrial and meteoritic rocks (Middlehurst, Burley, 1966)[1] thought comparable to the lunar surface materials seem to show that the solar corpuscular bombardment, the fluorescence due to the short-wave radiation of the Sun, and the thermo-luminescence induced in some rocks by heating, all fall sadly short of the energy requirements of the observed glows and bright spots. Nor is there any clear correlation with solar activity; in some cases, indeed, the correlation was found to be negative. The accelerations imparted to solar protons in the Earth's 'magnetic tail' might improve their chances of producing a glow of required intensity, but this would require the LTPs to occur preferentially near full phase, which is the opposite of the recorded incidence.[3, 18]

The explanation by meteoritic impacts could at best account for a glow lasting a few seconds, but not for one exceeding an hour in duration and of a pattern observed by Greenacre and Barr in 1963, which has been subsumed to be volcanic in the earlier part of this chapter. By their very nature the distribution of meteoritic impacts must be random, whereas the reported LTPs are heavily crowded about rayed craters, the edges of the maria and lunabase crater floors, where volcanic or ultravulcanian activity may be expected. Moreover, while no correlation with the events on the Sun has been established, the records of lunar events show a major peak near the perigee of the lunar orbit and a minor one near apogee. This corresponds exactly to the fluctuation in the body tides due to the Earth's varying attraction, which by disturbing the crustal fractures would affect volcanic activity and/or the escape of gas from the interior.

There is, however, one possible cause of luminescence that has so far been left out of consideration. Porous or fairy-castle materials, such as are to be found on the lunar surface, strongly sorb gases. Sorption increases with decreasing temperature, so that some of the gas imprisoned in the rocks during the night cold would be released with the return of the Sun, even as Pickering suggested, although he did not invoke this mechanism. Such desorption is accompanied by electrification, so that glow discharges may be expected.[9]

In fact, the sum total of the evidence presented above points to the existence on the Moon of a daytime 'atmospheric skin' of no great depth and density, but considerably more massive than conventional atmospheric models allow, as first proposed by Pickering (1903)[22] and seconded by myself (1959)[8, 9, 10] on somewhat different grounds. We shall reconsider the matter in greater detail further on, and will attend to the related problems of lunar colours and seasons first.

REFERENCES

1 BURLEY, J. M., and MIDDLEHURST, B. M. (1966). 'Apparent Lunar Activity: Historical Review', *P.N.A.S.U.S.*, **55**, p. 1,007 ff.
2 CAMERON, WINIFRED S. (1965). 'An Appeal for Observation of the Moon', *R.A.S.C. Jrnl.*, **59** (5), p. 219 f.
3 CAMERON, W. S., and GILHEANY, J. J. (1967). 'Operation Moon Blink and Report of Observations of Lunar Transient Phenomena', *Icarus*, **7** (1), p. 29 ff.
4 FIELDER, GILBERT (1961). *Structure of the Moon's Surface*. Pergamon Press, Oxford, Etc.
5 FIRSOFF, V. A. (1956). 'Lunar Occultations Observed in Blue Light and the Problem of the Moon's Atmosphere', *J.B.A.A.*, **66** (7), p. 257 ff.
6 FIRSOFF, V. A. (1958). British Patent No. 802, 427 (Date of Appln. 15 Sep. 1954).
7 FIRSOFF, V. A. (1958). 'Colour on the Moon'. *Sky and Telescope*, **17** (7), p. 329 ff.
8 FIRSOFF, V. A. (1959). *Science*, **130**, p. 1,337 f.
9 FIRSOFF, V. A. (1959). *Strange World of the Moon*. Hutchinson, London, and 1960, Basic Books, New York.
10 FIRSOFF, V. A. (1960). *Science*, **131**, p. 1,669 f.
11 FRANKLIN, F. A. (1967). 'Two-Color Photometry of the Earthshine', *Jrnl. of Geophysical Research*, **72** (11), p. 2,963 ff.
12 KOPAL, Z., and RACKHAM, T. (1964). 'Lunar Luminescence and Solar Flares', *Nature*, **202**, p. 171 ff.
13 KOPAL, ZDENEK (1966). 'Luminescence of the Moon and Solar Activity' in *The Nature of the Lunar Surface – Proceedings of the 1965 IAU–NASA Symposium* (Eds. W. N. Hess *et al.*). The Johns Hopkins Press, Baltimore, Md.
14 KOZYREV, N. A. (1959). *Sky and Telescope*, **18** (4), p. 184 ff.
15 KOZYREV, N. A. (1959). *Sky and Telescope*, **18** (10), p. 561.

16 Kuiper, G. P., and Middlehurst, B. M. (Eds.) (1961). *Planets and Satellites, The Solar System, Vol. III.* University of Chicago Press.

17 Middlehurst, B. M. (1966). *Chronological Listing of Lunar Events.* NASA Goddard Space Flight Center, Greenbelt, Maryland.

18 Middlehurst, B. M. (1967). 'Events on the Moon', p. 116 ff. in *1968 Yearbook of Astronomy* (Ed. Patrick Moore). Eyre and Spottiswoode, London.

19 Middlehurst, B. M., Burley, J. M., Moore, P., and Welther, B. L. (1968). *Chronological Catalog of Reported Lunar Events.* NASA TR R–277.

20 Moore, Patrick (1965). 'Evaluation of the Reported Lunar Changes', *Annals of the New York Academy of Sciences,* **123**, p. 797 ff.

21 Moore, P., and Cattermole, P. (1967). *The Craters of the Moon.* Lutterworth Press, London.

22 Pickering, W. H. (1903). *The Moon.* Doubleday, New York.

23 Robinson, J. H. (1968). 'Herodotus and Schröter's Valley', *J.B.A.A.,* **78** (2), p. 116 f.

24 *Sky and Telescope* (1968), **36** (5), p. 299 f.

25 Smith, A. G. (1953). *The Strolling Astronomer,* **7**.

26 Wright, W. H. (1929). 'The Moon as Photographed by Light of Different Colours', *P.A.S.P.,* **41** (241), p. 125 ff.

Abbreviations used

P.N.A.S.U.S. =Proceedings of the National Academy of Sciences of the U.S
R.A.S.C. Jrnl.=Journal of the Royal Astronomical Society of Canada.
J.B.A.A.   =Journal of the British Astronomical Association.
P.A.S.P.   =Publications of the Astronomical Society of the Pacific.

14

Colours, shadows and seasons

Lunar colours are among those matters on which people are inclined to hold opinions as strong as they are unreasonable. It was hoped that the colour photography from the *Apollo VIII* would dispel all doubts and hush the voices of dissent, but it has only served to make the confusion worse confounded.

Visually the Moon appeared to the crew to be wholly devoid of colour – 'greenish-brown', 'sandy-brown' or even plain grey. This is, of course, a correct enough description of the general first impression of the Moon's face as seen through a telescope from the Earth. Yet the simultaneous colour photographs seem to tell a very different story. In most of these the Moon looks incredibly green, ranging from khaki to grass and deep pine-green. One low-light picture is even more incredibly blue, indicating severe under-exposure (or is it Earthshine?). The colourings shown on the Earth, on the other hand, seem to be correct.[15] It looks as if the meter was set for the Earth, whose surface brightness exceeds that of the Moon about $5\frac{1}{2}$ times, whence the trouble. Nevertheless, we have a photograph of the large ring of Langrenus on the border of Mare Foecunditatis, where the crater itself and its inner halo are of a desert sandy brown hue, with perhaps a little touch of red or pink here and there, but the mare surface still looks green.

Which, then, gives a truer account of the situation, the eye or the camera?

Seen through a telescope under moderate magnification, the Moon is a dazzling sight. The dark-adapted eye is suddenly deluged with light, reflected from a surface that, though intrinsically dark (average albedo of 0·07), is in full sunlight. In these conditions any faint coloration tends to be wiped out and leave an impression of uniform monochrome. I have had a similar experience in the mountains, walking or climbing along a ridge in bright sunshine with a wide expanse of sky in front; when I lowered my eyes on to the scene at my feet it appeared like a black-and-white, or perhaps rather sepia, photographic print. Yet this was certainly an illusion.

The position of a spacecraft crew orbiting the Moon fits in somewhere between the two cases. As the astronaut Anders observed, 'The sky is very, very stark. The sky is pitch black and the Moon quite light. The contrast between the sky and the Moon is a vivid dark line'. In these circumstances the eye will not be a very good judge of colour; and various hues, though mostly faint, have been seen on the Moon by telescopic observers, at least some of these hues having been confirmed by the impartial test of filter photography. The latter include a touch of green in Mare Foecunditatis near Langrenus.

Colour photography is temperamental; correct exposure and processing are crucial for obtaining the right effects. The energy of light is inversely proportional to the wavelength. Thus the emulsion will generally register the violet and blue first and the red last. As a result under-exposure or under-development result in blue highlights, while penetration by light produces orange 'flashes', and over-exposed photographs tend to lose colour and the blues in particular. A green bias may be ascribed to slight under-exposure; grey-blue rocks do occasionally come out green, and this may apply to the Moon – in one of the greenest *Apollo* photographs the intermediate highlights are blue. Yet there is no valid reason why Langrenus should be a drab brown, while the adjoining mare is a deep jungle-green, in the same photograph, where any fault in exposure or processing would have affected both in the same way. The cover of the September 1968 issue of *Sky and Telescope* shows the new Hawaiian Mountaintop Observatory, and the colourings of the surrounding volcanic landscape are not unsimilar to those of the Moon in some of the *Apollo* pictures. As already intimated, Mare Foecunditatis gives a green reaction when a red or infrared filter is used, so that some green is definitely there. Basic rocks with a high olivine content are of just about the right shade.

The other fault of colour photography is that when one particular colouring predominates it tends to invade the neighbouring areas of a different colour; this, though, results in a loss, not excess, of colour.

Various attempts at straight telescopic colour photography, using Earth-based telescopes, have been equally discordant and inconclusive.[8, 21] Visual observation offers a much more delicate means of detecting colour, despite the disturbing intervention of our atmosphere. When the Moon is high and clear some colour can usually be made out among the general dun-brown monochrome, especially at full phase. One of the most definitely coloured parts of the surface is the yellowish area to the north-west of Aristarchus, already mentioned on p. 181 and often referred to as 'Wood's Spot' after its discoverer R. W. Wood, although, being some 200 miles wide, it is quite a 'spot'. There are similar yellow 'spots' about Sinus Roris. Almost equally well marked is the greenish khaki hue over a circular area where the Apennines link up with the Haemus Mountains. Palus Somnii is golden, and a bluish-green undertone is discernible in the dark 'spots' in the region of Sinus Aestuum and Mare Vaporum, particularly if the eye is withdrawn from the eyepiece to isolate them.

Some observers have reported 'vivid greens', mostly in the maria seen at small phase, but also, for instance, on the southern slopes of Maginus and near the Rheita Valley. H. P. Wilkins[24] notes that Mare Humorum is 'often a fine light green, especially close to sunrise and sunset; while the Sea of Peace (Mare Tranquillitatis) is a clear greenish tint near full moon'. Apparently it can look very green when 'the sky is very clear and the lighting favourable'. I have never seen it thus myself, at most just a faint touch of olive. But I have found a yellowish-green cast to be widespread on the Moon, which yields some substance to the *Apollo* colourings. Rudaux and de Vaucouleurs have compiled a map of lunar colours, which appears in the Larousse encyclopaedia *Astronomie*.[20] Its rainbow effect is far beyond anything that could be seen on the Moon at any one time; but it must be appreciated that the colours may vary with the angle of illumination, so that they do not appear there all at once, quite apart from any transient glows described in the foregoing chapter. Moreover, the map lacks a neutral background, which leaves the colourings suspended in the air and throws them into excessive relief. When all this has been taken into account, the representation seems to be accurate enough and in good agreement with the findings of most other observers, including myself; although some, D. P. Avigliano[3] for one, have been unable to see any of the greens and blues or violets, and found only yellow, a little red and grey. This, he himself suggests, may be a matter of colour sensitivity, or perhaps colour appreciation, in which many people are remarkably deficient.

When all is said and done, however, lunar colourings remain faint. The light of the Moon is much redder than the direct sunlight, and according to B. Lyot is best matched by a mixture of grey and brown volcanic ash. In average conditions the telescopic view of the Moon does not reveal any colour at all, although the differences of albedo, or *shade* and *tint* – which two terms denote the result of adding black and white respectively to any given colour – may be strongly marked. Such subtle chromatic gradations as there exist are best winkled out by the use of filters.

This can be visual or photographic. The filters employed in ordinary photography, however, are not really suitable for this purpose, whence the comparative lack of success of the earlier visual students of the Moon who had nothing better at their disposal. W. H. Pickering[16] used a blue glass filter in visual observation towards the end of the nineteenth century, while filter photography of the Moon appears to have been pioneered by R. W. Wood of 'Wood's Spot' in the U.S.A. in 1910, although some work on similar lines was initiated more or less simultaneously in Germany and Russia.[9] Good monochromatic filters, however, had not become readily available until after World War II, and, curiously enough, their visual application encountered some bigoted opposition from the old school of observers.[12]

Broadly speaking there are two types of filter that can be used to advantage in colour research : the narrow-band monochromatic filter and the tricolour separation sets of three filters. A narrowband filter eliminates all wavelengths of light outside its *passband*, which may vary in width, and thus isolates a substantially pure spectral colour. This necessarily entails some loss of brightness and so requires a relatively large aperture or low power; but this limitation has been grossly exaggerated. The incidental advantage of such a monochromatic filter is that it effectively suppresses all residual chromatic aberration, as well as some of the boiling, and so yields a sharper image. The late Director of the Meudon Observatory, A. Danjon, recommended using a neutral green filter for this purpose.[7]

In a tricolour separation set the red, green and blue filter respectively isolate not pure spectral colours, but as nearly as possible the responses of the three types of cone in the retina which determine the colour sensation (Figs. 16 and 17).[18] The band of transmission is much wider, with some overlap in the wings of the

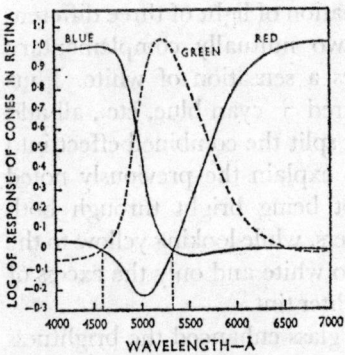

Fig. 16. Response of the three different types of cones in the human retina to different wavelengths of light (after Pirenne).

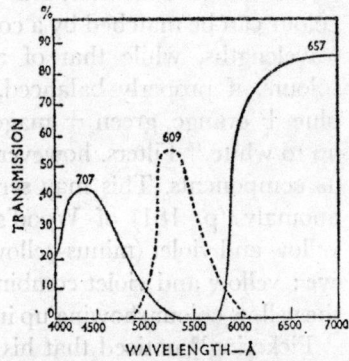

Fig. 17. Transmission characteristics of Dufay tri-colour filter set.

respective curves, though there is a strongly marked peak. In most cases there is not a great deal of difference between the tricolour red and green and the corresponding monochromatic filters, which simply intensify the effect, but there may be some discrepancy at the blue end of the spectrum.

A colour analysis by means of filters operates on the principle of suppression. A red monochromatic filter will suppress all light except the red within its passband. Thus any green, blue or yellow object, if its colour is spectrally pure, will look black, and a red one more or less bright depending on its shade, or albedo. In reality, though, pure spectral colours are of rare occurrence, so that all we can expect is a change in the scale of intensities. If, therefore, we compare, say, a view of a crater in blue and red and it looks brighter in red, it means that the crater is redder than its surroundings. In a tricolour set a yellow area will be equally bright in red and green, a green one will be dark in red, and a violet one in both. The positive image always corresponds to that in the complementary colour : in other words, a red filter will give a minus-red, or peacock blue, distribution of colours, and only its negative will show the true incidence of the red tints.

A filter analysis, however, goes beyond what the eye can see, not only because it can be extended into the infrared and ultraviolet, either photographically or visually by means of an electronic image converter, but because any visual impression of colour can be matched by a combination of light of three different wavelengths, while that of any two mutually complementary colours, if properly balanced, gives a sensation of white. Thus blue + orange, green + magenta, red + cyan blue, etc., all add up to white.[18] Filters, however, will split the combined effect into its components. This may serve to explain the previously noted anomaly (p. 181) of Wood's Spot being bright through both yellow and violet (minus-yellow) filters, while looking yellow to the eye : yellow and violet combining to white and only the excess of the yellow colour showing up in a lighter tint.

Pickering[16] noticed that his blue glass enhanced the brightness of the limb and of the polar regions, as well as some crater rays and haloes. The earliest systematic study of lunar colours by filter analysis, however, was made in Germany by A. Miethe and B. Seegert in 1911.[10] They projected on to a screen diapositives obtained with an ultraviolet and an orange filter through screens of complementary colour, which by reversing the density scales gave the correct chromatic effect. In this way they were able to identify various colorations of the lunar surface, and most of these were subsequently confirmed at Lick by W. H. Wright and Dorothy Applegate in 1926.[25]

They used the 36-inch Crossley Reflector (reflectors are superior to refractors for colour work owing to the absence of chromatic aberration) to photograph the Moon through ultraviolet, violet, green, orange, red and extreme red filters, and obtained a number of fine negatives, which leave no doubt of the objective existence of many colourings previously observed and recorded by Miethe and Seegert.

Wright presented these findings in a paper accompanied by an original photographic print, showing three doublets of the Moon at the age of 10, 15 (full) and 20 days respectively taken at 3,600 Å (ultraviolet) and 7,600 Å (extreme red).[25] The comparison is highly instructive. As Wright observes, the differences between the two sets of images are not great, but local inequalities are unmistakable. There is generally more light of the longer wavelengths in the terrae, including the rayed craters, while the maria appear somewhat smoother and darker in the ultraviolet, which he does not mention. Since the maria are also dark in the extreme red, except for the central portion of Mare Serenitatis, the northeast of Mare Imbrium, Mare Frigoris and parts of Oceanus Procellarum, this suggests a greenish cast. The darkest minus-red areas, comparatively pale in the ultraviolet, are in Mare Crisium, parts of Foecunditatis, Tranquillitatis, round the edges of Serenitatis, Imbrium near Sinus Iridum, the dark 'spots' near the centre of the disk, and west of Kepler in Oceanus Procellarum. The great darkness of Grimaldi in longer wavelengths found by Miethe and Seegert is not confirmed. There are some differences between the phases, the maria becoming more uniform towards the full in ultraviolet, as well as more differentiated in extreme red. The photographs are too small to show minor detail, but it is clear that considerable differences exist in the terrae as well, the floors of the large rings in the Ptolemaean highlands in particular being darkish in the ultraviolet exposures, where the Fourth Quadrant is scattered with a multitude of bright dots.

A very different method was used by a British colour photographer N. W. Scott (1964).[21] Having failed to obtain a reliable colour picture of the Moon by direct means, he resorted to 'colour correcting masks', i.e. film diapositives having about 70 p.c. of the contrast shown in the negatives, which when 'applied to high-contrast negatives made from suitable transparencies yielded, on colour paper, clear local variations in colour for different parts of the lunar surface'. These variations reappeared at successive phases and lunations, as well as being in good agreement with the earlier results, which puts their reality beyond question. Colour enhancements of 50 – 100 times were obtained in this way.

I have before me a combined red-and-blue print of a waxing gibbous Moon made by this method. The contrasts are very harsh, but generally concordant with the gentler gradations of the

Wright-Applegate photographs, as is further demonstrated by a pair of the latter printed in superimposition in colour. There are some differences, however, which seem to be attributable to phase, full in the second case, where the blue-green element in the central and eastern maria is much stronger than in Scott's own photograph of the gibbous Moon.

Details apart, the presence of a green undertone in Mare Foecunditatis is confirmed. Among the areas showing a red reaction Scott himself lists the outer rays and 'interspersed material' of Copernicus, the northern rays of Kepler, the floors of Plato and of the walled plains around Alphonsus. The dark spots within the latter are described by him as 'brownish', and the dark areas in the region of Sinus Aestuum and Mare Vaporum as 'blue'. In addition, the photograph shows the redness of the Moon near the terminator, also remarked upon by Avigliano,[3] and an intense blue band along the Sunward limb, which, due allowance being made for the colour enhancement, confirms Pickering's earlier report.

A very interesting observation made by Scott was of colour variations in the integrated light of the Moon, which shifted from a 'sandy yellow' to a 'silvery blue' within a matter of 24 hours or so. The effect was confirmed with a tricolour photometer. The absolute (unenhanced by processing) increase in the blue component was often up to fourfold; there was a parallel intensification in the ultraviolet. I have observed blue enhancements as well, but ascribed them to terrestrial atmospheric causes. Apparently, however, there is no such correlation, and we may recall that a blue enhancement was recently recorded by Hynek with the highly sophisticated apparatus at the Corralitos Observatory (p. 186), which, while corroborating Scott's finding, also casts some doubts on the efficiency of that sophistication. Scott has failed to discover any correlation between the 'blue Moons' and solar activity, in which respect these would resemble the behaviour of the LTPs.

In 1964 E. A. Whitaker[23] took some photographs of the Moon with the McDonald Observatory's 82-inch reflector through an ultraviolet filter of about 3,800 Å and an infrared one of 7,800 Å. In printing a positive transparency of the infrared photograph was combined with the ultraviolet negative – a method similar to Miethe and Seegert's – adjusting the contrast of the positive, to eliminate so far as possible all differences of albedo. This technique yields a picture where the differences of shade represent true colour differences, the lighter areas being spectrally the bluer and the darker the redder. These differences are quite strongly marked, in general agreement with the results obtained by other methods. Whitaker suggests that the distinctively coloured plateaux in the maria, e.g. in the central part of Mare Serenitatis, correspond to lava or ash flows.

We thus have clear instrumental records of colour differences on the lunar surface, although some of these may lie in the ultraviolet and infrared parts of the spectrum, inaccessible to the eye. Yet, while such results are very valuable, the eye is an extremely sensitive detector, and visual observation is in many respects superior to photography. Nor is it at all true that observers are too subjective in their judgement : I rather suspect that this is comparatively rare. On the other hand, instrumental techniques, to take colour photography as an example, are not wholly free from 'whims' either. Moreover, the tests just described cover the extreme ends of the colour spectrum and miss the intermediate hues.

Most distinguished selenographers, to mention only Gruithuisen, Mädler, Klein, Schmidt, Birt, Pickering, Wilkins and Haas,[9] have seen some colour on the Moon; but the first systematic visual study of the problem was carried out in the years 1934–37 by the German *Arbeitsgruppe für Mondbeobachtungen* (Working Group for Lunar Observations) under the direction of F. Kaiser.[9] Among other colourings they reported some green in Mare Serenitatis, occasional brown in Plato, and an intriguing seasonal cycle in Ptolemaeus, whose floor changed from grey in the morning to olive green and to yellowish in the evening, simulating the growth and decay of vegetation. These observations, however, do not appear to have been confirmed, or perhaps – repeated.

A similar concerted effort to determine lunar colourings was made in the U.S.A. in 1936–38 by a group headed by W. H. Haas,[9] who thus sums up its results :

2. Low-sun colours are relatively striking but very transient, seldom being distinguishable between relative colongitudes 30° and 150° (in other words, within the central 2/3 of the lunation

– see p. 202). The hues are usually greens, sometimes browns, and perhaps occasionally purples (or blues). Though infrequent even along the terminator, these colours affect chiefly very dark crater floors near that line.

3. High-sun colours are very inconspicuous but very stable . . . The colours are blues (or purples) in the high-sun dark areas, especially the darkest ones; browns in delicate darker shadings on dark floors, the use of different filters having confirmed these browns to the extent of showing that some marks are redder than others; and greens over large dark areas, although greens more than 2 days from the terminator are very rare. Blue and brown are probably sometimes both present in the same area. (P. 43, Ref. 9).

The latter remark refers to the colouring which Rudaux and de Vaucouleurs describe as *enfumé* (smoked).[20] The filters used in this work were neither monochromatic nor tricolour separation ones, which made them purely ancillary to direct visual colour estimates. This differentiates it from my own investigation of lunar colourings extending over the three years from 1954 to 1956 systematically and afterwards sporadically. During that earlier period I observed the Moon at every phase and every lunation, whenever the weather obliged, primarily by means of a Dufay tricolour separation set of filters, mounted in a peripherally bayed rotatable disk between the eyepiece and the object glass or mirror. Some hundreds of colour determinations were made, and their results summarized in the May 1958 issue of *Sky and Telescope*.[5]

Thus in viewing the Moon successively through different filters I have found that : in *red* (tricolour) the maria look darker and the surface relief is clear at the terminator, the differences in brightness due to the angle of illumination being largely smoothed out and the white-light brightness of the limb subdued; in *yellow* (monochromatic) the maria are much paler and all contrasts are reduced, except for some minor dark spots, such as the 'pseudo-shadows' of Eratosthenes (see p. 205), which Pickering[16, 17] has described as violet, and which stand out dark and clear; in *green* (tricolour) the bright markings, and the rays of Tycho in particular, are enhanced, and so too is the brightness of the limb, but the terminator detail is darkened; In *blue* (tricolour) the terminator

Plate 18. The surroundings of Riccioli with lava flows and suspected signs of glaciation (compare with Plate 12a). Note also the graben-rilles crossing the walled plain (IV–173H₁). *N.A.S.A.*

(a)

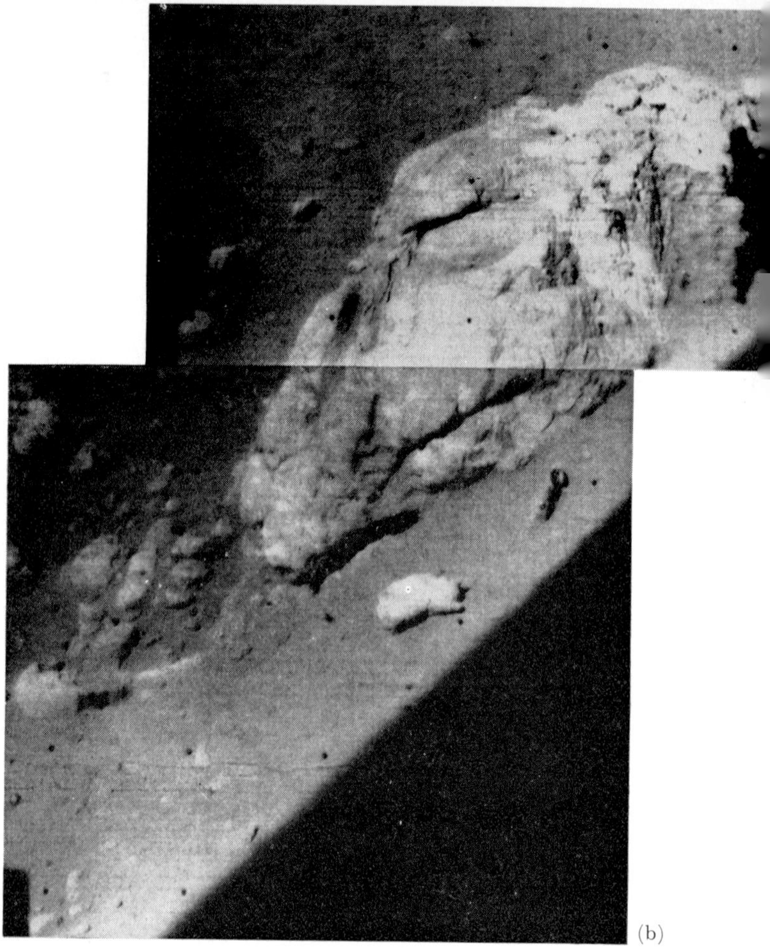

(b)

Plate 19a. A 'reindeer moss' of rock. A *Luna-9* close-up of the lunar surface as received at Jodrell Bank and built up by facsimile photography by London Express Pictures (page 236).   *London Express.*

Plate 19b. A volcanic bomb? 45-cm boulder photographed by *Surveyor I* (p. 236).   *U.S. Information Service, London.*

Plate 19c. Scree on the outer slopes of Tycho, a *Surveyor-VII* picture.
  U.S. Information Service, London.

Plate 19d. An unusual type of crater photographed on the far side by *Apollo X.* It is very deep and almost precipitously steep on the inside, but the bounding ring wall is very low and gentle on the outside; the interior is filled with irregular mounds which seem to have been formed by landslides. Such a crater could be produced by a powerful underground explosion, and be of meteoritic origin.

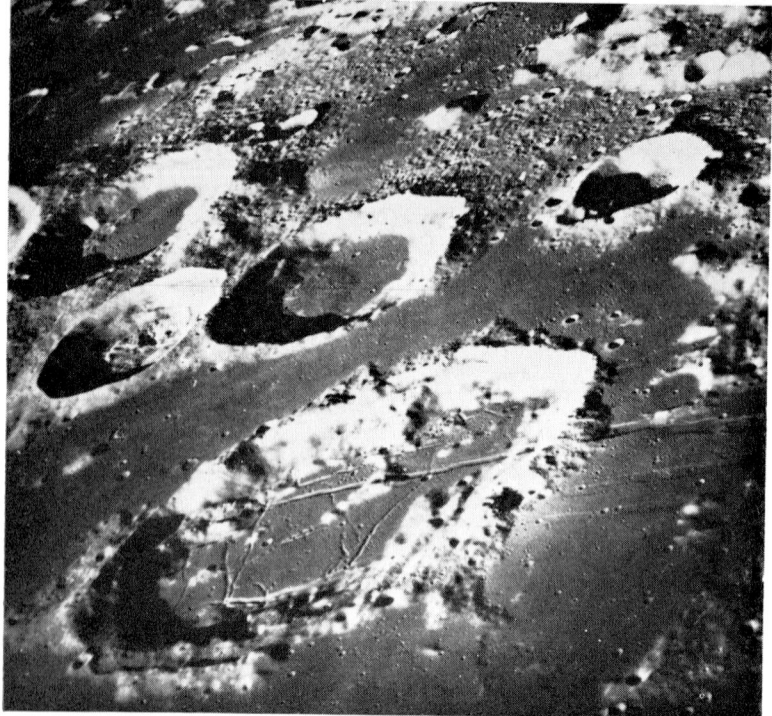

Plate 20a. Looking over the sunset terminator into Mare Tranquillitatis. A fine crystal haze seems to be reflecting the low Sun in the upper central part of the picture (p. 223) (V–52M).

N.A.S.A.

Plate 20b. An *Apollo-8* photograph of the mountain ring of Goclenius and its neighbours, illustrating the variability of the depth/diameter ratio. The deep-seated character of graben-rilles is clearly apparent.

U.S. Information Service, London.

lights are very dull, but the brilliance of the limb and Pickering's 'polar caps' (p. 196) is greatly increased, as also is that of some bright areas associated with rayed and haloed craters; In *violet* (monochromatic) the general aspect of the Moon is dull, bar exceptionally good seeing, the limb brightening is still there, but thinner than in blue, while the 'polar caps' and some other blue-bright features appear slightly hazy.

That the blue-violet limb brightness is not simply a matter of surface colouring may be gauged from the fact that the lunabase valleys near Mare Orientale, when revealed by favourable libration, are at times nearly invisible with a blue or violet filter, although quite clear in other colours. This may be taken to correspond to Scott's 'blue-Moon condition', but also parallels D. Alter's[2] finding that when Alphonsus and Arzachel were photographed by him in 1955 in the light of different colours with the 60-inch reflector at Mt. Wilson 'For some reason the blue-violet photographs lost more detail in the west (now east) side of Alphonsus than they did in the floor of Arzachel. This is not true of the infra-red ones . . .' He resisted, not very convincingly, the temptation to regard this as evidence of a local atmosphere, but it was this observation that induced Kozyrev to pay special attention to Alphonsus and eventually detect gas emission there (p. 178).[14]

The colourings in the maria seen by me are substantially the same as those revealed photographically, but I have also found systematic seasonal shifts of colour. Thus the Moon as a whole is greener at lunar sunrise than at sunset, and greener at crescent phase or near the terminator than under steeper lighting. Although the Earthshine is blue, it stands out clearest through a green filter, which is again an indication of a green element in the colouring of the Moon – in this context at night. The weakening of the green component towards noon is paralleled by an intensification in the violet chromatic component in the maria, some craters and a few special localities. When seen through a violet filter, the maria look much paler and less differentiated at full, an effect that is also noticeable in Wright's ultraviolet photographs. Some colours, however, are unaffected by lunar seasons and probably represent the genuine colorations of the surface.

Reproduced here (Fig. 18) are the colours mapped by me in Grimaldi and Plato by reference to an intensity chart of various colours seen with the same filters. There are also some faint greenish markings, almost invisible without filters, such as the radial bands in Copernicus. These are seasonal and, like the violet-tinted dark areas in Eratosthenes, appear only under a high Sun, being seemingly variable withal, at least in the latter case.[17]

Since the lunar equator is tilted to the plane of the ecliptic at a constant angle of barely $1\frac{1}{2}$ degrees, the Moon can have no seasons other than the times of its $29\frac{1}{2}$-our-days-long day, except just around the poles. The situation is the exact opposite to that

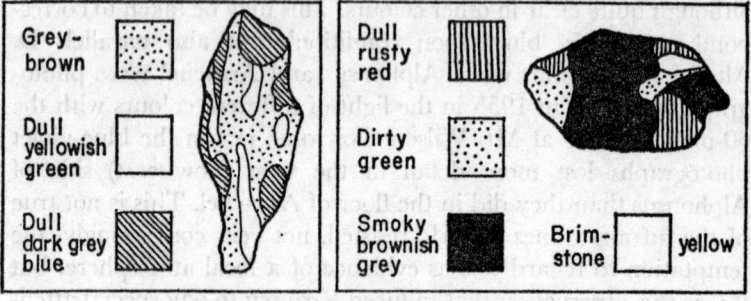

Fig. 18. Colour distributions in Grimaldi (left) and Plato (right), seen with a $6\frac{1}{2}$-inch reflector at 200x, from 22:30 to 22:50 Universal time, 6th August 1955. The boundaries are only approximate.

of Mars. As seen from the Earth, Mars stays substantially full all the time, so that its appearance is not visibly affected by the 24-hour axial period, while the equatorial obliquity of 24° results in seasons comparable to ours and even more strongly marked owing to the doubled length of the year. Thus any changes observed on the surface of Mars are directly attributable to the seasons. On the Moon, however, the climatic cycle coincides with the lunation, which is the lunar day. Thus the morning is also the spring, midday – high summer, evening – autumn, and night – winter. The seasons are coincident with the phases and move with the shadows.

The appearance of any given area may alter drastically with the solar colongitude, which is numerically equal to the selenographic longitude of the morning terminator measured eastwards from the mean centre of the disk, being approximately 270° at new, 0° at first quarter, 90° at full, and 180° at last quarter. It is the arcuate distance of the sunrise line from the centre of the disk measured in the sense of rotation round the lunar globe. Thus, whereas we can say with certitude that the Martian maria darken in the summer owing to climatic factors, a very similar darkening of the lunar maria is usually held to be due to the angle of illumination alone. Variations in their colouring may be a more conclusive indication of thermal effect, but even colours can change with the angle of incidence and reflection. Nevertheless, this is one of the reasons why the believers in the dogma of Dead Moon are so dead-set against lunar colours.

Dogmas apart, there is not much difference between a change in the colour, on the one hand, and in the shade or tint, on the other, except that the latter may be readily ascribed to lighting, but *may* is not the same as *must*, and it is often very difficult to decide whether such a change is real or merely apparent. As already noted, the maria darken towards the full phase and grow paler again as the Sun westers in the lunar sky. The lunabase crater floors and other dark features generally share this behaviour, while the bright markings pursue the opposite course.[6, 9] Not often is much thought spared for this monthly cycle, and yet it has some puzzling aspects. Owing to the rough and porous texture of the lunar surface, a waxing half-Moon supplies only about 8 p.c. of the light of a full Moon, so that the changes in the overall surface brightness are dramatic, and it may seem reasonable at first sight that all contrast of shade and tint should increase towards full. Yet the illumination of a mountainside is at its peak not when the Sun is directly overhead, but when its rays fall at right angles to the slope. This factor is invoked to explain the apparent brightness of the limb, where the hillsides facing the Sun when this is behind the observer become crowded by perspective. How this is supposed to account for a preferential brightening in blue light is another question.

However, let us observe the waning crescent when the Moon is between 23 and 25 days old. The western terrae look very bright, possibly owing to that crowding of Earthward hillsides, but the maria are very pale, much paler than at full phase or at dichotomy. Moreover, it seems to make a difference whether the light comes from the east or from the west at the same angle, for the maria appear darker at a waxing than at an equivalent waning phase. Indeed, it is well known that a waning crescent is brighter than a waxing one, although there are more maria in the west and the terrae are brighter than the maria in both cases. The objection that they are not the same maria is unconvincing anyway, for they all look equally dark at full, and the western maria stay dark till after the last quarter. This is part of the general diurnal asymmetry, the maria and most dark features being darker in a waning gibbous than in a waxing gibbous Moon. Nor is this a matter of casual impression : this conclusion is sustained, for instance, by precise photometric measurements of calibrated photographs of Plato, whose floor is darker in the afternoon than at an angularly corresponding hour in the morning.[22]

Moreover, not all maria darken simultaneously. This is particularly true of Mare Frigoris, the most polar of the maria, and the northernmost part of Oceanus Procellarum, which tend to stay pale after the other maria have gone dark. Mare Frigoris is sometimes so pale as to be barely distinguishable from the neighbouring uplands; nor is it consistent in this respect, and its shade appears to vary from lunation to lunation. This is not a matter of libration either, for the Mare may be at its palest when geometrically farthest from the limb. We would thus seem to have here not just a seasonal change, but something like weather. I would not press this point too hard, as the effect has never been properly investigated, but it deserves to be.

There is diurnal asymmetry of the behaviour of some crater shadows as well that is difficult to explain in geometrical terms.[4, 9] Haas notes that often a recently shadowed area stays darkened for some days after the Sun has reached it, without any apparent relation to the angle of slope. Temporary darkenings and brightenings could be due to crystalline material giving a concerted reflection at a particular angle of illumination. But it is difficult to see why dark markings should emerge under a high Sun when all shadows have gone, and they do. These are the so-called 'pseudo-shadows', for they seem to be wholly divorced from the configuration of the ground.

Probably the best known case is that of Eratosthenes, which is well placed for observation near the centre of the disk. As the Sun climbs up into the lunar sky and the last chink of shadow quits the wide bowl of the mountain ring, a system of dusky smudges, lines and spots begin to develop, to become conspicuous at noon. These markings climb indiscriminately over walls and terraces and pay no respect to the known relief. Pickering, who had spent years studying them under the clear skies of the mountains of Jamaica, maintained that the distribution of the 'pseudo-shadows' was subject to capricious changes, and this led him to suggest that they were due to 'swarms of small animals' crawling about the crater.[17] Low-trailing volcanic vapours would be a more plausible explanation, but the movement of the markings itself has been denied by some observers, whether on sufficient evidence is another matter.

The *dark radial bands*[1, 14] which spring into prominence in many craters (188 such are listed by L. J. Robinson – 1963),[19] mostly those with bright rays and haloes, under a high Sun, may be placed in the same category as pseudo-shadows. Patrick Moore[14] has seen some of the bands in Aristarchus dissolve into smaller spots when observed with the great Meudon refractor, and suggests that they may consist of chains of dark-haloed craterlets. This, though, finds no confirmation in the *Orbiter* photographs. The bands of Aristarchus also appear to alter from day to day, not wholly by reason of seeing. The *Orbiter* photographs, taken under oblique lighting, are wholly inexplicit in this respect. The less-conspicuous 'Maltese-cross' type of marking, found in my filter observations in such craters as Copernicus and Bullialdus,[5, 6] is probably of a similar origin.

Bright features, too, largely vanish under oblique illumination and do not achieve full prominence until full, or a little after. In fact, Jan van Diggelen has found that the maximum brightness is normally delayed by up to 18 hours. These features seem also to be subject to diurnal asymmetry in being generally larger and more prominent in the morning than in the evening, which is a time of reduced contrast.[8] It may be that if the bright and dark matter is concentrated at the bottom of small hollows it will require a high angle of illumination to become visible. This is the usual explanation of the bright rays not becoming visible until the Sun is 8 – 10° above the horizon (Pickering, Lenham). It is further corroborated by Thornton's observation that some rays of Tycho look dark under lower angles of lighting.[11] On the other hand, lunar close-ups show bouldery material in minor crater haloes.

At full phase the rays near the limb stand out as clear as those at the centre of the disk, which may be accounted for by the hollows casting no visible shadows when the Sun is directly behind the observer. But this does not explain why the rays continue to be seen at sunset under lower angles of illumination than at sunrise, for surely the hollows cannot be preferentially oriented. This is well illustrated by the Mt. Wilson pair of photographs No. 256, the second whereof shows a 26-day-old Moon. In this the rays from Olbers can be seen right on the terminator. If, however, we take, for instance, the Lick D 507 photograph of a 4-day-old waxing crescent the rays of Proclus, usually brighter than those of Olbers, are invisible to the west of the crater, which is some 300 miles from the terminator, and barely discernible on Mare Crisium in the east. Even the halo of Langrenus is extremely pale.

It seems that the visibility of the rays depends not just on the angle of illumination but also on the length of exposure to sunshine; in other words, the effect is at least partly thermal.

Contrariwise, some of the bright haloes appear to be smaller and paler in the evening than in the morning. Micrometric measurements of the bright aureole round Linné made by Barnard, Pickering, Wirtz and Haas support this view.[9, 16] Barnard used the 40-inch Yerkes and 36-inch Lick refractors to measure the Linné halo before and after a lunar eclipse and found an increase. These results have been denied by other investigators,[13] but, their state of mind being prejudiced, this was perhaps to be expected.

Bright spots have a way of materializing in such rings as Posidonius and Gassendi (Plate 17) seemingly out of nowhere, in a way difficult to account for by illumination alone. Some kind of activity appears to be involved. Perhaps Pickering[16] was right, after all, in contending that 'lunar vulcanism' does not begin to manifest itself till 2 or 3 days after sunset, if 'vulcanism' is equated with the emission of gas and the formation of deposits.

To sum up, the evidence of the last two chapters shows that the Moon is a scene not only of sporadic transient activity, which appears to be wholly or largely endogenic, but also of parallel seasonal changes, in which some of the former may be only a local intensification. There are also large-scale fluctuations in the seasonal cycle, comparable to our weather. All this points to the existence of at least temporary daytime atmosphere of far greater density and mass than is conventionally admitted.

REFERENCES

1 ABINERI, K. W., and LENHAM, A. P. (1955). 'Lunar Banded Craters', *J.B.A.A.*, **65** (4).

2 ALTER, D. (1955). 'A Suspected Partial Obstruction of the Floor of Alphonsus', *P.A.S.P.*, **69** p. 158.

3 AVIGLIANO, D. P. (1951). 'Lunar Colors', *The Strolling Astronomer*, **8** (5 & 6).

4 AVIGLIANO, D. P. (1952). 'Unusual Lunar Shadows', *The Strolling Astronomer*, **7**.

5 FIRSOFF, V. A. (1958). 'Color On the Moon', *Sky and Telescope*, **17** (7), p. 329 ff.

6 FIRSOFF, V. A. (1959). *Strange World of the Moon*. Hutchinson, London.

7 FIRSOFF, V. A. (1961). 'The Visual Observations of the Planets with Colour Filters. *Discovery*, **22** (6), p. 231 ff.

8 FIRSOFF, V. A. (1965). *Surface of the Moon*, Hutchinson, London.

9 HAAS, W. H. (1942). 'Does Anything Ever Happen on the Moon?', *Jrnl. of R.A.S.C.*, **36** (7).

10 MIETHE, A., and SEEGERT, B. (1911). *Astronomische Nachrichten*, **188**, p. 371 ff.

11 MOORE, PATRICK (1954). *Guide to the Moon*. Eyre & Spottiswoode, London.

12 MOORE, PATRICK, *et al.* (1964). 'Colour Filter Report', *J.B.A.A.*, **74** (4).

13 MOORE, PATRICK (1965). 'An Evaluation of the Reported Lunar Changes'. *Annals of the New York Academy of Sciences*, **123**, p. 797 ff.

14 MOORE, P., and CATTERMOLE, P. (1967). *The Craters of the Moon*. Lutterworth Press, London.

15 N.A.S.A. (1969). *Apollo VIII – Man Around the Moon*, EP 66.

16 PICKERING, W. H. (1903). *The Moon*. Doubleday, New York.

17 PICKERING, W. H. (1919 – 25). 'Eratosthenes, 1 – 6'. *Popular Astronomy*. Nos. 269, 287, 295, 312, 317.

18 PIRENNE, M. H. (1948). *Vision and the Eye*. Pilot Press, London.

19 ROBINSON, L. J. (1963). 'Banded Craters', *J.B.A.A.*, **73** (1), p. 33 ff.

20 RUDAUX, L., and DE VAUCOULEURS, G. (1948). *Astronomie*. Larousse, Paris.

21 SCOTT, N. W. (1964). 'Colour on the Moon', *Nature*, **204** (4963), p. 1,075 f.

22 *Sky and Telescope* (1959). **19** (2), p. 22.

23 WHITAKER, E. A. (1966). 'The Surface of the Moon' in *The Nature of the Lunar Surface – Proceedings of the 1965 IAU – NASA Symposium* (Eds. W. N. Hess *et al.*), p. 79 ff. The Johns Hopkins Press, Baltimore, Md.

24 WILKINS, H. P. (1954). *Our Moon*. Muller, London.
25 WRIGHT, W. H. (1929). 'The Moon as Photographed by Light of Different Colors', *P.A.S.P.*, **41** (241).

Abbreviations used

J.B.A.A.   =Journal of the British Astronomical Association.
P.A.S.P.   =Publications of the Astronomical Society of the Pacific.
Jrnl. of R.A.S.C.=Journal of the Royal Astronomical Society of Canada.
N.A.S.A.   =National Aeronautics and Space Administration.

15

For and against a lunar atmosphere

The study of the Moon has little to do with psychology, but as an organized human pursuit it is a social activity just as much as politics or gardening. The ostensibly scientific arguments may send long taproots into the subconscious mind, where often enough lies hidden a superstitious fear, almost a hatred, of the idea of life on other worlds, especially the near ones. Thus the belief in the airlessness of the Moon comes to be held with irrational tenacity, and any challenge to it, however mild, meets with fierce resistance, as though it were an outrage to common sense. Closely, if perhaps needlessly, allied to this is an uncritical adulation of mathematical theories, which, in addition to their snob value as academic status symbols, are invested with quasi-magical powers – *mathematica locuta causa finita*. In sober truth, however, a mathematical theory is merely an artificial logical construct, valid only within the assumptions made at the start. In the case of straightforward relationships it is comparatively easy to devise a system of simplifying assumptions amenable to mathematical treatment and adequate in practice. Yet, as the situation grows in complexity, this system of assumptions bears less and less resemblance to reality, and the whole procedure degenerates into an abstract intellectual game, possibly not entirely meaningless, but suggestive rather than definitive, and at times, neither, as we have already seen. Observation and experiment remain as the ultimate touchstones of truth, but not infrequently what is passed for observational or experimental findings, or 'results', is not the actual data, but their interpretations in the light of some theory or other, so that we are back to where we started from.

In thinking of the lunar atmosphere we must consider the past and the present condition of the Moon, the latter being largely a fruit of the former.

We have seen in Chapter 11 that the sinuous rilles must be identified as ancient watercourses of a form peculiar to an extremely arid climate, with intense evaporation, and having their origin in underground springs, seemingly associated with volcanic activity, rather than in atmospheric precipitation. The existence of small lakes at the head of the streams seems unmistakable, but there is evidence for more extensive water basins, probably of no great depth, and local glaciation as well.

All this demands a very appreciable atmosphere. A ground-level barometric pressure of 20 mmHg was suggested (p. 148). This corresponds to a boiling point of about 22° C, a temperature that was certainly exceeded at midday in low lunar latitudes, where most of these river-beds are found, so that the water would literally boil away. And yet some of the channels are over 200 km long, so that if anything the estimated pressure may be too low.

We have no certain way of dating this selenological episode, and there have most probably been several such, following on the formation of a mare or other orogenic and vulcanic disturbance. Mare Orientale, however, marks one such period, which is recent in lunar history (p. 161) and could hardly be more than 500 million years old. The 'Seasons-of-the-Year' maria girdling Mare Orientale on the Earthward side are contemporaneous or subsequent to it, and here and there thin but long streams have meandered down over them from the neighbouring mountains. The Moon must, therefore, have possessed an adequate atmosphere for some time after the formation of the Orientale complex. It quite certainly has no such atmosphere at the present moment, but it follows that an airless condition is by no means axiomatic for bodies of lunar size.

We may recall that J. J. Gilvarry[11] went further than this and proposed in 1960 that the maria of the Moon were at one time filled with water to an average depth of 2 km, and that it would require a 'period of the order of 10^9 years' for an atmosphere based on such a hydrosphere to be lost to space by molecular dissipation. Water is dissociated by the solar ultraviolet into its constituent hydrogen and oxygen. Hydrogen, being the lightest of all gases, would be lost quite quickly; in fact, our own atmosphere contains very little hydrogen. The dissipation of oxygen, which is 16 times heavier, will take much longer. So long, however, as there is water both gases will continue to be generated, the relative amount of oxygen rising and yielding with a proportion of other gases an atmosphere suitable for life 'as we know it'. At the same time, as the short ultraviolet splits the diatomic oxygen molecules, the free oxygen atoms thus produced will combine with the residual molecules at a certain atmospheric level to the triatomic state, O_3, and an ozonosphere will form, filtering out the ultraviolet rays harmful to organic molecules and arresting the further photo-dissociation of water. This will be additionally frozen out of the upper atmosphere at the cold trap, being thus doubly shielded from loss to space. Gilvarry does not appear to have included these favourable factors in his calculations, but they further strengthen his case.

There exists no present evidence that the situation envisaged by him ever existed, though some areas of the maria may have been flooded at one time; but the reasoning of the preceding paragraph is equally applicable to the more modest atmospheric episodes corresponding to the sinuous rilles.

These problems cannot be clearly understood without some knowledge of the theory of dissipation of planetary atmospheres, a classical treatment of which was given by Sir James Jeans in 1925[15] (*See* Appendix, pp. 248 ff.).

In an *ideal gas* there is no cohesion between perfectly elastic molecules, which collide and rebound in all directions, tending towards the condition of *equipartition of energy,* where the kinetic energy of every particle is the same. Since kinetic energy is a half of the product of mass into the square of velocity, equipartition demands the ideal velocity of a particle to be inversely proportional to the square root of its mass. If the respective particles are a molecule of hydrogen and oxygen, which is – as stated – 16 times heavier, the hydrogen molecule will be moving 4 times faster. Absolute temperature provides a measure of molecular agitation and the square of the theoretical equipartition velocity as defined above, which is known as *mean square-root* molecular velocity, is directly proportional to the absolute temperature, which becomes zero when the mean square-root molecular

velocity drops to zero. This occurs at the temperature of
$-273\cdot1°$ C, known as *Absolute Zero,* from which absolute temperature is measured upwards in °K (Kelvin), which are the same
as °C, save for the point of origin. The absolute temperature is
always positive and equal to the centigrade temperature
$+273\cdot1°$ K.

Real gases approximate to this idealized behaviour at moderate
pressures and temperatures above the *critical temperature,* which
is specific for every gas and below but not above which it can be
liquefied by pressure alone. At lower temperatures a gas becomes
a *vapour*; its molecules grow sticky and tend to lump up temporarily into larger aggregates. For any given temperature any
particular vapour has a *partial pressure* (i.e. that part of the total
amospheric pressure which is due to the proportion of the vapour
in the mixture) at which it is in equilibrium with the liquid or solid
phase. If this pressure is exceeded condensation will generally
occur until the partial pressure is brought back to its equilibrium
value; although if the atmosphere is undisturbed, electrically
neutral and free from suspensions (aerosols) the excess vapour
may stay *uncondensed* in the unstable *supercooled condition.* The
temperature at which condensation should occur with a given
partial pressure, or concentration of the vapour, is known as the
dew point. At atmospheric level where the temperature is
sufficiently low for most of the vapour to condense and be substantially eliminated from the supernatant atmosphere is called
the *cold trap.*

The critical temperature of water is $374°$ C $= 647°$ K, so that
at the temperatures normally encountered on or near the lunar
surface steam will be a vapour, and even at $60°$ C its molecules are
largely combined into threesomes, or *trihydrols,* which raises
their weight from 18 (atoms of hydrogen) to 54. The critical temperatures of some common gases are: Argon, $-122°$ C; Carbon
Dioxide, $31\cdot1°$ C; Hydrogen Sulphide (H_2S), $100°$ C; Nitrogen,
$-147°$ C; Oxygen, $-118°$ C; Ozone, $-12\cdot1°$ C; Sulphur
Dioxide (SO_2), $155\cdot4°$ C. Since midnight temperatures on the
Moon may drop below $-150°$ C, even nitrogen may become a
vapour by night.

The equipartition of energy, however, is a purely statistical
concept, portraying the average result of the movements of very
large numbers of molecules. This is the same as saying that the

actual molecular velocities are scattered on both sides of the mean square-root molecular velocity appropriate to their weight and the absolute temperature of the gas, and now and again a molecule may attain a velocity equal to or greater than the velocity of escape from the atmosphere. If, therefore, this velocity were directed away from the planet and the path of the molecule were unobstructed it would never return, and clearly the lighter and the hotter the gas the more often this would happen (p. 211).

At standard temperature and pressure, or *STP* for short, equal to $0°$ C $= 273\cdot1°$ K and 760 mmHg respectively, a cubic centimetre of any gas contains $2\cdot69 \times 10^{19}$ molecules (Loschmidt's Number), and in these conditions the free path of an air molecule averages $9\cdot5 \times 10^{-6}$ cm. As the temperature rises and the pressure drops, however, the molecules become spaced farther and farther apart, and near the outer boundary of the atmosphere, at the *escape level* collisions become so infrequent that they can be neglected in practice (or is it in *theory*?) and those molecules whose outward velocity exceeds the velocity of escape will be lost to space.

For the purpose of mathematical treatment it is necessary to assume that the escape level is an interface, with gas on the one side and vacuum on the other. This is not really true, as, on the one hand, the atmosphere thins out upwards gradually into the interplanetary gas, and, on the other, at the distance of the Earth from the Sun this gas has an average density of 1,000 particles per cm^3, these particles being mostly protons and electrons, but including some heavier atoms and dust grains.

However, Jeans considers that immediately below the escape level the gas may be regarded as though it were contained in a vessel open at the upper end and any molecule moving faster than the velocity of escape and at such an angle as to emerge through the aperture would escape. The velocity of a molecule may be represented as a vectorial sum of three components : one vertical and two at right angles to it and to each other. The rate of escape is obtained by integrating these components over the appropriate range (a fuller mathematical treatment will be found in the Appendix, pp. 248 ff.). It may then be taken that the instantaneous rate of escape remains unchanged and corresponds to the equivalent loss of mass from a layer 1 cm thick at the base of the atmosphere. Jeans, further, assumes that the atmosphere is *isothermal,* i.e. has the same temperature throughout, and takes a fictitious height which it would have if its density were throughout the same as at the base. Obviously, when the time required for a complete centimetre layer at the base to be lost to space is multiplied into this height expressed in centimetres the entire atmosphere of a given gas will have gone.

Now, if we have a mixture of gases every gas in it will have a different mean square-root velocity, inversely proportional to the square root of its particle (atom, molecule or higher aggregate) weight. If the mixture is uniform the lighter constituents will be lost at a higher rate, and so the atmosphere may be regarded as a sum of partial atmospheres of its components, and the rate of dissipation of each partial atmosphere computed separately. This is, no doubt, very handy, but at the top of the atmosphere gases tend to become stratified according to their particle weight, and no molecule can escape unless it is physically present at the escape level. The Earth and Venus have hydrogen haloes, and hydrogen does escape, but very little of the other atmospheric constituents. Moreover, if the constituent is a vapour it will not obey the Gas Laws, and may be substantially frozen out of the atmosphere above the cold trap, so that again it will be very inadequately represented at the escape level. We have seen that all the likely gases will be present on the Moon in the condition of vapour during the night and in some cases through most of the day as well, especially in the polar regions. On the other hand, chemical compounds may be broken up into their elementary constituents by photo-dissociation, and, for instance, the escape of water to space must be considered in the terms of atomic hydrogen and oxygen and not of water molecules as such, which again raises various complications.

Jeans's original analysis of the situation is defective in several respects, in addition to the difficulties already mentioned.

To begin with, he made an error, for the time he gives for the complete loss of the atmosphere of a given gas corresponds only to a drop in its density to $\frac{1}{e}$ of the original value, where e is the base of natural logarithms approximately equal to 2·72. Even more misleading is the assumption that the atmosphere is isothermal.[18]

The real structure of the atmosphere of a terrestrial planet, such as the Earth or Mars, and presumably the Moon during a 'water period', is quite different. The lower atmosphere, or *troposphere,* which in the case of our air comprises 4/5 of its mass and extends up to about a quarter of its height – if the molecular spray of the *exosphere* is disregarded, is in a state of *convective equilibrium,* ensured by ascending and descending currents. This approximates to the *adiabatic condition,* where there is no net loss or gain of heat and the changes of temperature are due solely to those of pressure (and so volume). Such an atmosphere is thoroughly mixed in its gaseous constituents, though the presence of vapours, such as water vapour, subject to condensation and vaporization, serves to complicate matters. In an atmosphere thus constituted the temperature decreases upwards at a steady rate, known as *adiabatic lapse,* which depends on the composition and gravity and for dry air on the Earth amounts to 6° C per km. The troposphere terminates upwards in the *tropopause,* as the boundary dividing it from the supernatant *stratosphere* is called. The lower part of the stratosphere is substantially isothermal and extends in the case of the Earth for about 30 km. Higher up the temperature rises again owing to the presence of ozone, which becomes heated by the absorption of ultraviolet rays, and to ionization.

In sum total the troposphere is much denser than a corresponding isothermal atmosphere would be. On the other hand, ionization makes the temperature of the escape level much higher than assumed by Jeans, this being put in the case of the Earth at between 1,000 and 2,000° K.

Lyman Spitzer, Jr. (1952)[18] has amended Jeans's escape times by taking these two points into consideration and introducing a correction, based on the ratio of the *particle density* (number of particles per cm^3) at the escape level, put for the Earth at an altitude of 150 km, to that at the base of the atmosphere. The correction is a function of the temperature of the escape level. If this is taken as 500° K, Jeans's escape times must be multiplied into 6.9×10^4; for 2,000° K, the correction factor becomes 9.7×10^5. Gilvarry's[11] calculations were based on Spitzer's formula, but he also took into account the existence of the hydrosphere, which had been previously overlooked, on the assumption that all the volatiles originally occluded in the lunar magmas had been released to the lunar surface at one go. This is not really possible, and we have seen from Kozyrev's observation of an eruption in Alphonsus that the emission of gases from the lunar interior continues to the present day. Gilvarry's graph is reproduced in Fig. 19.

There is, however, a further fundamental point that has eluded Jeans and Spitzer alike. They both regard the rate of escape as constant. Yet only the most energetic molecules are able to escape, while the less energetic are left behind. Molecular energy represents heat, and so with every escape some of the energy of the gas

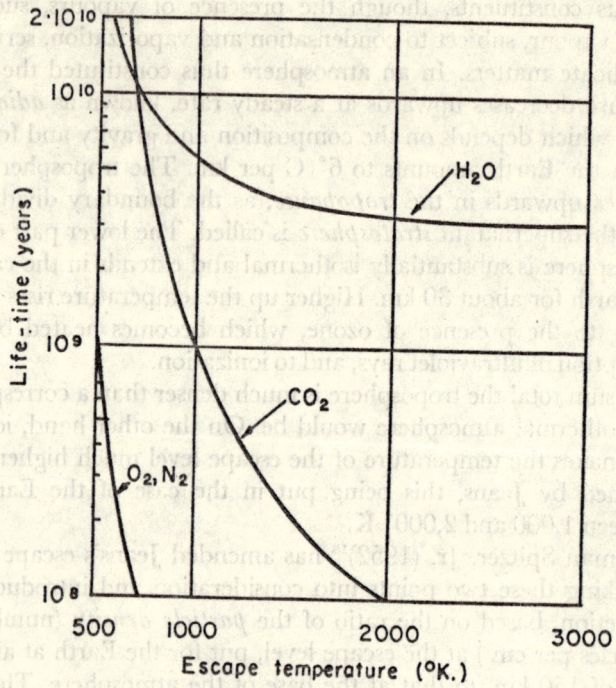

Fig. 19. Life-time of various constituents of the primitive lunar atmosphere and hydrosphere, as a function of assumed temperature in the escape layer. The curves for oxygen and nitrogen are indistinguishable on the scale used.

or its original supply of heat in the gas is lost to space and its temperature must inevitably go on dropping until it becomes so low that all molecular dissipation has effectively ceased (*See* Appendix, p. 251). It is a self-refrigerating process. In these circumstances the high temperatures attributed to the escape level are not credible.

Ionization is the source of these temperatures and it is, therefore, held to represent a threat to the existence of a gas envelope round bodies of small mass. There may be some truth in that, but it is not all one-way traffic, and much will depend on what gas is ionized.

In ionization electrons are split off the atoms and form a kind of extremely light gas, having a particle weight 1/1,840 of a hydrogen atom (or proton). If this happens at the escape level such a gas will be lost to space almost instantly, and will continue to diffuse away differentially until the growing excess of positive ions has built up a sufficient electric charge to make the combined electrostatic and gravitational attraction acting upon an electron equal to the force of gravity exerted upon a positive ion less the repelling force of the positive charge also acting upon it. Once this has happened, electrons and ions will escape to space at the same rate and the charge will stop growing. In the case of hydrogen this condition is reached when the positive electrostatic force becomes equal to half of the gravitational force (van de Hulst[19] – see Appendix, p. 251). But if the gas is singly-ionized argon, whose atomic weight is 40, it still carries the same charge as a proton, while the gravitational force exerted upon it is 40 times as great; consequently, an electrostatic retaining force which simultaneously repels positive ions, corresponding to 20 gravities will be needed to achieve the condition of escape equilibrium. The lunar gravity may be only $\frac{1}{6}$ g, but even so the resulting electrostatic attraction will be equal to 3·3 g, or more than the surface gravity of Jupiter!

The effect of such an electrostatic charge will be twofold: it will immediately induce a corresponding negative charge in the subjacent layer of gas, or if near the surface – in the ground, to which the positively-charged gas will thus become electrostatically bound;[8, 10] the attraction of this negative charge will cause the positively charged layer to become compressed. Although the range of the positive electrostatic force will be short, this is bound to have a profound effect on the rate of escape not only of argon, but of other molecules, even if neutral, as molecular collisions within the escape layer will become more frequent with an increase in its density. Moreover, a strong positive charge at the escape level will repel, or at least slow down, the high-energy protons emitted by the Sun, and so act as a kind of invisible armour shielding the underlying atmosphere from being swept away by the solar wind.

This is no special pleading, but plain straightforward physics.

It is the enthusiasts of the dead-Moon idea who are guilty of distortion, which step by step snowballs into absurdity (*See* Appendix, pp. 252 ff.).

To sum up, while molecular evaporation of gases to space is a physical reality, and it is true that bodies of small mass are generally devoid of atmosphere owing to it, our present mathematical apparatus for estimating its long-term effects is wholly inadequate.

Jeans himself has calculated that if the mean square-root molecular velocity of a gas exceeds a quarter of the velocity of escape the whole of this gas will be lost within about 1,000 years; but should it drop to a fifth of the velocity of escape the 'lifetime' of the gas would be of the order of one aeon. He, therefore, concluded that the Moon could not retain oxygen, nitrogen or water vapour for 'cosmic periods' ($>10^9$ years), but ought to be able to hold on to gases with molecular weights of 25 and over.[15] On this basis Kuiper[18] assumed in 1949 that carbon dioxide (CO_2) of m.w. 40, carbonyl sulphide (COS) of m.w. 60, ozone (O_3) of m.w. 48, and sulphur dioxide (SO_2) of m.w. 64, the latter of which is produced by volcanic action and meteoric impact alike, might be among the spectroscopically detectable constituents of a lunar atmosphere. He searched for ozone and sulphur dioxide, but failed to find either, and concluded that if their absorptions corresponded to the partial pressures of 0˙005 mmHg for ozone and 0˙0003 mmHg for SO_2 respectively at STP they could not have escaped detection. Since, though, it is completely unrealistic to assume a barometric pressure of one atmosphere on the Moon, and under lower pressures the absorption lines become much thinner and correspondingly more difficult to see, this finding is not particularly meaningful. Another eligible gas is argon, which, however, has no absorption within the accessible spectrum. Argon is formed by the radioactive decay of potassium-40, and Harlow Shapley has calculated that if the Moon had retained all of the argon thus produced over the geological time it would possess an atmosphere of this gas equal to about 10^{-3} of ours in ground density. This calculation disregards not only possible loss to space, but also the effect of sorption, and in any case most of the radiogenic argon would stay imprisoned in the rocks.

Like argon, nitrogen is inaccessible to Earth-based spectroscopy, and any gas whose absorptions are blanketed by telluric lines would be equally beyond reach. CO_2 forms barely 0.03 p.c. of the terrestrial air by weight. If it were transferred to the Moon, it would yield a partial pressure of 0·04mmHg, or just over 1/20,000 of an atmosphere. Not only is this rather more than is at present thought possible, but under the low barometric pressure obtaining on the Moon the resulting thin absorptions would be completely masked by the telluric CO_2 and could not be detected. On the other hand, we shall recall (p. 178) that during the solar eclipse in 1879 some Australian astronomers observed changes in the spectrum of the Sun close to the lunar limb. The Astronomer Royal Airy noticed an 'unusual absorption' in yellow in the light of the eclipsed Moon on the night of 23rd – 24th August 1878, and we have Kozyrev's spectrograms, showing a number of emission lines in Alphonsus in November 1958. He also reported in 1961 emission lines of hydrogen in Aristarchus. He did not seem to think the hydrogen could have come from the photo-dissociation of water. This observation has been questioned, which does not prove anything either.

Spectral analysis is not a very sensitive test for the presence of an atmosphere. For instance, the Galilean satellites of Jupiter had long been thought to be devoid of gaseous cover, despite some secondary indications to the contrary, but in 1963 distinctive absorptions were discovered in their spectra at the Crimean Astrophysical Observatory (Kalinyak).[16] The issue of the lunar atmosphere could be decided spectroscopically by observations from space or, of course, from the lunar surface itself. None such, however, have been made at the time of writing.

Another classical method of detecting and estimating a planetary atmosphere is by observing the behaviour of stars at occultation. When an atmosphered body passes between the star and the observer the light of the star will be dimmed, reddened and refracted at the limb. The small angular displacement of a star due to the refraction in the Moon's putative thin atmosphere is difficult to measure. However, Comstock and Pickering observed widely-spaced double stars close to the limb of the Moon and claimed to have found relative shifts in their positions of between 0·2″ and 0·4″, whence an atmosphere of about 1/8,000 of our air in ground density was inferred.[13, 24, 25] These observations have been discounted, but I have no record of their having been checked up and disproved.

According to textbooks a star snaps out instantly at contact with the lunar limb, although it has occasionally been reported to linger. This is usually attributed to the irregularities of the limb profile. Refraction should make the star visible for a while after it has actually been covered up by the Moon and to reappear a little before it has been uncovered. The time of the occultation would thus be slightly shorter than obtained from theory. The effect to be expected is very small, and once more the presence of mountains and valleys on the lunar limb makes sufficiently accurate prediction impossible. If, however, long series of observations are examined statistically the deviations due to the irregularities on the lunar limb should cancel out, and the intervention of an atmosphere would make the occultation diameter of the Moon somewhat short of its micrometric measure. If, though, there is obscuration at the limb as well this will act the opposite way, extending the time of occultation, so that the test is double-edged.

Investigations of this type have been made. In 1864 the occultation diameter of the Moon was found at the Royal Greenwich Observatory to be 4″ less than its measured value. The *American Ephemeris* for 1925 gave a smaller discrepancy of 1·5″. A refractive shift of 1″ would correspond to a lunar atmosphere 0·001 of ours in density, and figures between 1/2,000 and 1/750 were obtained in this way at various observatories.[13] The tendency is to dismiss such estimates because of their date, on the assumption that we know better today; this may be true as a general proposition, but the point is that no similar observations have been made in recent years, and excessive reliance may have been placed on less direct methods.

Refraction is inversely proportional to the wavelength and scattering by gas to the fourth power of the wavelength. It, therefore, occurred to me that an occultation could be observed with a binocular eyepiece simultaneously through a red and blue, or infrared and ultraviolet filter, either visually or photo-electrically, to see if any difference would appear. E. A. Whitaker,[28] then of the Royal Greenwich Observatory, thought this was a promising method and was planning to try it out, but other things supervened and it all came to nothing. He suggested, however, that there might be a 'blue flash' at occultation, analogous to the green flash occasionally seen immediately after sunset in the denser atmosphere of the Earth. With this for inspiration I observed the occultations of two stars (ι Tauri, $4^{m}\cdot7$, and 150 Tauri, $8^{m}\cdot5$) on 14th March 1957 with a $6\frac{1}{2}$-inch reflector, using a blue-violet filter. Neither star snapped out as expected at contact with the Earth-lit limb of the Moon, but dimmed rapidly, then flashed out in seemingly undiminished brightness, to dim again and vanish. The Moon was $2\frac{1}{2}$ days old at the time, so that its dark limb was close to the evening terminator, which may be an important point, for when I repeated the same observation with an older Moon no comparable effect was found.[7, 9] So far as I know no attempt has ever been made to check up this result in the conditions corresponding to those of my original observation.

There are other similar effects that have not been fully investigated. When the Moon occults Jupiter or Saturn a dark band has often been seen on the disk of the planet along the bright, though not the dark, limb of the Moon. Barnard, Douglass and Pickering succeeded in photographing this dark band, which formed a steep angle with the belts on the planets' disks, precluding confusion with them. As measured on Pickering's negatives the dark band was $3''$ wide. Its absence at the dark limb was explained by the absence of an obscuring layer during the lunar night. Pickering also claimed to have measured a flattening of the disk of Jupiter, non-coincident with its natural oblateness and equal to $0\cdot5''$.[25] Like most of his observations, this, too, was dismissed as unreliable, while the dark-band effect was explained away as a contrast effect. The source of the effect, however, is in the brain and should not affect the photographic plate, and it may be worth recalling that Pickering was right about his 'riverbeds', after all, improbable as these may have seemed at the time.

During partial eclipses of the Sun bright points of light have been repeatedly seen on the dimmed limb of the Sun at contact with the Moon. I saw them, too, in 1954 and 1959. The usual explanation by 'symmetrically spaced solar flares' will not do, quite simply because, as the Sun and the Moon move, the bright points do not disappear, but merely change position on the solar limb. Irradiation seems also to be ruled out, as I observed the Sun in projection on to a white card, where the surface brightness of the image was low, and, moreover, I reduced it deliberately as a further check.[9] I strongly suspect that the 'bright points' can be seen at every eclipse. Diffraction might be responsible for this effect, but it could not be called in to account for the 'dark band'

as well, because it would affect the image of the planet equally at
the bright and the dark limb of the Moon ... And we must not
forget that Australian observation of absorptions in the solar spectrum at the lunar limb (p. 178).

There are indeed other indications that a lunar atmosphere of
perceptible density exists. The fact that the moonlight is redder
than sunlight may be ascribed to the intrinsic colouring of the
lunar surface, although the reddening of the reflected light is
normally regarded as a possible indication of filtering by gas. On
the other hand, the intensification of the red element towards the
terminator, recorded both visually and photographically, cannot
be explained by intrinsic colouring. Nor can the apparent intensification of the violet element in the maria with the progress of the
lunation which becomes readily understandable on the assumption that some gas is released from the surface in the heat of sunrays, possibly supporting a fine haze of crystals or dust. A bright
limb is usually due to a layer of gas, which is thickest in the line
of sight at the limb. The Earth, Mars and Venus are bright at the
limb from this cause; see, for instance, the second photograph of
the Earth on p. 12 of *Apollo VIII* pamphlet.[23] Now it has already
been mentioned that this effect is present on the Moon as well. It
may be due to some extent to the fact that the highest points of the
lunar surface are also the brightest (p. 65), and so in becoming
crowded towards the limb run together into a bright line, but this
cannot account for the differential brightening-up in blue light,
again recorded both visually and photographically (p. 201). This
limb brightness appears clearly in the *Apollo VIII* colour photographs. The colours may be distorted, but this is a separate problem; and if we accept the brightness of the limb in the colour
photographs of the Earth taken on the same film with the same
camera, it would be rather illogical to deny its reality in the case of
the Moon. The colour photograph of the Moon on p. 14 of the
Apollo VIII pamphlet is unacceptably green, but it shows a bright
line all along the limb, bordered inwards by a bluish haze. Fielder
says that 'the telescopic view of Mars indicates that surface markings become indistinct as the limb it approached : no such loss of
detail is apparent on the Moon',[6] which is correct in integrated
light, but when a blue filter is used there is a noticeable loss of
detail[9] (on Mars, of course, the surface becomes more or less

invisible through a blue filter). This loss of detail is clearly shown in the colour photograph mentioned above.

It is equally clear in another *Apollo VIII* photograph, reproduced *inter alia* on the cover and p. 12 of the pamphlet. In the original there is further a kind of haze or sheen below the Earth, recalling the 'Sun-haze' seen from a high point on a frosty morning over the surface of the Earth. An *Orbiter V* frame 52M (Plate 20A), giving an oblique view of Mare Tranquillitatis across the evening terminator into the sunlit hemisphere, seems to display this effect even more strongly; as though the rays of the low Sun were scattered back by a haze of fine crystals over the lunar surface. Finally, Astronaut Lovell noticed a 'fan-shaped fine white haze radiating out from behind the Moon's horizon about two minutes before the Sun comes up over it'.[5] Flying 100 miles above the surface, he would be looking obliquely through the atmosphere, partly illuminated by the still-invisible Sun, and this would give a fan effect. However, this haze and/or the supporting atmosphere appears to be too thin to be seen in full sunlight, but then neither are the stars.

All this tallies well with the observations detailed above, including my own of lunar occultations in blue light.

Of course, if there is gas exhalation from the interior, as the various transient phenomena (Chapter 13) indicate, an atmospheric halo round the Moon must be expected. Further evidence in support of its existence comes, oddly enough, from Gold's hypothesis of lunar dust (p. 173). His conception is physically sound: silicates do darken and are pulverized when exposed to short-wave and corpuscular radiation of the Sun, so that a dark dust layer should form on the lunar surface if this is unshielded against such radiation. Yet there is no dark dust there, in fact – no dust at all: a jet of nitrogen gas repeatedly directed at the lunar ground along a leg of the *Surveyor I* from a distance of 15 cm failed to produce any disturbance of the surface. Moreover, the trenching operations by the *Surveyor III* have revealed that the mare material is brighter on the exposed surface than underneath, where its grains have a kind of 'lunar varnish', similar to that on our desert rocks.[17, 27] Thus it would seem that the destructive radiations of the Sun do not reach the lunar surface, or do not reach it in unabated strength, as they would in a vacuum.

Before concluding the 'case for the defence', we must consider three further points: lunar meteors, aurorae and twilight.

In 1921 J. W. Gordon[12] estimated that about 20,000 meteorites should strike the dark side of the Moon on an average observing night, producing telescopically observable flashes. None such, however, appear to have been seen. This could be explained by the smaller meteoroids being burned out by atmospheric friction before they had a chance of hitting the Moon, which requires surprisingly little gas. In a conventional isothermal model, under a reduced gravity, the density of the lunar atmosphere would decrease much more slowly with height than on the Earth, and would still be adequate to bring a meteorite to incandescence and produce a meteor trail if its ground density were only 10^{-7} of ours. Meteors attaining in our skies magnitudes between -6 and -8 would be within range of a medium-sized telescope against the unlit hemisphere or off the limb. A search for these was instituted by the American Association of Lunar and Planetary Observers between 1941 and 1946 and about a score of reports of lunar meteors, distinguishable by their relative faintness, slow motion and shortness of the trails, had been received by the end of this period.[14] Some of these reports had come from L. La Paz, E. K. White and W. H. Haas, all of them experienced astronomers, but little has been heard about this since.

Even less gas is needed for auroral phenomena and daylight aurorae have been observed on the Earth at a height of 1,000 km. The search for lunar aurorae has been generally unavailing, although I did observe something like an auroral streamer near the south pole of the Moon on 24th May 1955.[9] The effect may have been due to instrumental causes. On the other hand, the Moon appears to be substantially devoid of polar magnetism, so that polar lights would be unlikely there in any case, while the various glows seen on the Moon may be partly similar phenomena, associated with local magnetization or electric fields. The *Surveyor VII* picked up some magnetic material on the slopes of Tycho.[22]

According to Russell, Dugan and Stewart[26] twilight should be observable on the Moon with a ground density of 0·0001 of our sea-level value, and, in fact, it has been repeatedly reported (Haas, Wilkins, Barcroft, Vaugn).[13] I have seen something like twilight myself and a kind of dilute brilliance round the dark limb of a very young Moon in a clear sky,[9] but am nevertheless sceptical on two counts: strong Earthshine can be easily confused with twilight; large objects, such as high mountains, can also scatter light and produce a twilight effect, and there are some huge peaks near the poles, especially the south, where the extension of the cusps, attributed to atmospheric twilight, has been seen.

In theory, polarization should allow us to isolate genuine twilight, and Fessenkov relied on it in 1943 in his search for lunar twilight on the dark side of the terminator close to the centre of a half-lit Moon, but found no effect, whence he inferred a density of less than 10^{-6} of our air. In 1949, however, another Russian astronomer, Yu. N. Lipskij, anounced the discovery of a lunar atmosphere equal to 10^{-4} of ours in ground density by the same method, but using a green filter with a passband centred on 5,300 Å, to eliminate the Earthshine.[6] The accuracy of this determination was called into doubt by Bernard Lyot and Audoin Dollfus in France,[21] who searched for the twilight effect photographically above the cusps on the dark side of a Moon at first quarter, using the 20-cm coronograph at the Pic du Midi Observatory, in 1949 and found no effect. This led them to fix the possible upper limit to the density of the lunar atmosphere at 10^{-8} of the terrestrial. The observation was repeated by Dollfus in 1950 with a more refined technique and yielded a still lower limit of 10^{-9}.[2]

A highly skilled man though Dollfus undoubtedly is, his method had certain inherent weaknesses. Thus he was looking for carbon dioxide on the dark side of the morning terminator at about 150 km above the surface, near the pole. At this time the ground temperature would be below $-150°$ C, and the atmosphere at this height may be colder still, so that any carbon dioxide it might contain would be largely frozen out. A Wratten 12 filter was employed, to cut out the background light scattered by the air. Now this filter cuts off all of the blue and violet part of the spectrum. The suspected lunar atmosphere would consist of very pure gas, probably monatomic and would, therefore, be intensely blue by comparison with our air, which even at the height of Pic du Midi contains various aerosols and scatters strongly the light of longer wavelengths passed by the filter. This being so, a positive result could hardly be expected.[7] Dollfus himself conceded these points in a private communication.

The next step was the investigation of electronic refraction of radio waves during occultations of radio sources at Cambridge by

Elsmore and Whitfield in 1955,[3] and Costain, Elsmore and Whitfield in 1956,[1] the latter of which resulted in fixing a new ground density for the lunar atmosphere at 10^{-13} of our sea-level value. This result from an occultation of the Crab Nebula is often given as just another upper limit,[6] which is incorrect, as a positive effect was observed, corresponding to an electron density of between 10^3 and 10^4 per cubic centimetre, and on the assumption of an isothermal atmosphere in the condition of rapid escape (after Link, 1956) yielded a ground-density figure of 'about 5×10^{-13} of the Earth's atmosphere at normal temperature and pressure' (Elsmore, 1957[4]).

We have seen, however, that in the absence of a magnetic field the Moon would lose most of the electrons in its exosphere to space, so that any direct comparison with the Earth, such as is assumed in this argument, is misleading. Rapid escape implies rapid chilling; the whole atmospheric model used in this connection is highly questionable anyway; and as Dollfus remarks the quoted result is 'somewhat hypothetical'.[20] Nevertheless, it is positive : it shows that the Moon has an atmospheric halo, however tenuous. Moreover, if we assume that the atmosphere of the Moon is due solely to the gravitational concentration of interplanetary gas[8], which 1,000 km above the surface has a mean particle density of $10^3/cm^3$ and a mean particle weight of 1, then the ground particle density of an isothermal atmosphere at $273°$ K so constituted comes up to only 10^4, which is about 0.01 of the value inferred from the occultation of the Crab Nebula.

This may be said to conclude 'the evidence for the Crown', which shows that the Moon is not wholly devoid of a high atmosphere, though its actual mass remains in doubt and Elsmore's figure may be regarded at most as the lower limit. Moreover, we must not forget Lovell's 'haze', which looks exactly like the kind of twilight Dollfus was looking for and failed to find, and this would mean a density of appreciably over 10^{-9} of the terrestrial even on a conventional view.

The existence of a low atmosphere, in a sense discontinuous with the upper levels and of a much higher density than either an isothermal or an adiabatic model permit, is strongly indicated by many other observations. Indeed, there are good physical reasons to think that these models are inapplicable to the present lunar conditions, which seem to favour evolution of gas from the surface in the heat of the Sun and its partial reabsorption in the night's cold.[9] We may soon have the complete answer. Meanwhile, this conception and the related issues of climate and erosion will be given fuller consideration in the next chapter.

REFERENCES

1 COSTAIN, C. H., ELSMORE, B., and WHITFIELD, G. R. (1956). *M.N.R.A.S.*, **116** (4).

2 DOLLFUS, AUDOIN (1952). 'Nouvelle Recherche d'une Atmosphère au Voisinage de la Lune', *C.R.*, **234**.

3 ELSMORE, B., and WHITFIELD, G. R. (1955). *Nature*, **176** (4479), p. 457 f.

4 ELSMORE, B. (1957). 'Radio Observations of the Lunar Atmosphere', *Philosophical Magazine*, **2**, p. 1,040 ff.

5 FAULKNER, ALEX (1968). 'Around the Stark, Lonely, Grey Moon', *The Daily Telegraph*, 31st December.

6 FIELDER, GILBERT (1961). *Structure of the Lunar Surface*. Pergamon Press, Oxford.

7 FIRSOFF, V. A. (1956). 'Lunar Occultations Observed in Blue Light and the Problems of the Moon's Atmosphere', *J.B.A.A.*, **66** (7).

8 FIRSOFF, V. A. (1959). *Science*, **130**, p. 1,337 f.

9 FIRSOFF, V. A. (1959). *Strange World of the Moon*. Hutchinson, London.

10 FIRSOFF, V. A. (1960). *Science*, **131** (3414), p. 1,669 f.

11 GILVARRY, J. J. (1960). 'Origin and Nature of the Lunar Surface Features', *Nature*, **188** (4754), p. 886 f.

12 GORDON, J. W. (1921). 'Meteors on the Moon', *Nature*, **107**, p. 234 f.

13 HAAS, W. H. (1942). 'Does Anything Ever Happen on the Moon?', *Jrnl. R.A.S.C.*, **36** (7).

14 HAAS, W. H. (1947). 'A Report on Searches for Possible Lunar Meteors', *Popular Astronomy*, **55**, p. 266 ff.

15 JEANS, J. H. (1925). *The Dynamic Theory of Gases*. Cambridge University Press.

16 KALINYAK, A. A. (1965). *Astronomiceskij Zurnal*, **42** (5), p. 1,067 ff.

17 KOPAL, ZDENEK (1968). *Exploration of the Moon by Spacecraft*. Oliver & Boyd, Edinburgh and London.

18 KUIPER, G. P. (Ed.) (1952). *The Atmosphere of the Earth and Planets*. 2nd Edn. University of Chicago Press.

19 KUIPER, G. P. (Ed) (1954). *The Sun*, p. 306 ff. University of Chicago Press.

20 KUIPER, G. P., and MIDDLEHURST, B. (Eds). (1961). *Planets and Satellites, The Solar System*, Vol. III. University of Chicago Press.

21 LYOT, B., and DOLLFUS, A. (1949). 'Recherche d'un Atmosphère au Voisinage de la Lune', *C.R.*, **229**.

22 NASA (1968). *Technical Report 32–1264.*

23 NASA (1969). *Apollo 8 – Man Around the Moon.* EP–66.

24 PICKERING, W. H. (1903). *The Moon.* Doubleday, New York.

25 PICKERING, W. H. (1925). 'The Lunar Atmosphere', *Popular Astronomy,* **35.**

26 RUSSELL, H. N., DUGAN, R. S., and STEWART, J. Q. (1945). *Astronomy.* Ginn & Co., Boston, U.S.A.

27 STUBBS, PETER (1967). 'The Shifting Sands of the Moon', *New Scientist,* **35** (556), p. 241 ff.

28 WHITAKER, E. A. (1956). Private communication.

Abbreviations used

M.N.R.A.S. =Monthly Notices of the Royal Astronomical Society.
C.R. =Comptes Rendus de l'Académie des Sciences.
J.B.A.A. =Journal of the British Astronomical Association.
Jrnl. R.A.S.C.=Journal of the Royal Astronomical Society of Canada.

16

Surface conditions

The fact that the lunar celestial pole keeps an angular distance of 1° 32′ from the pole of the ecliptic in the sky (p. 27) deprives the Moon of seasons and climatic zones comparable to those of the Earth and Mars. Only quite close to the poles can such a small axial inclination affect the climate to any extent. By the same token the Sun can never be directly overhead except within a narrow belt, 3° 4′ wide, centred on the lunar equator. In other words, the 'temperate zones of the Moon' extend over 173° 52′ of latitude.

As the Moon moves together with the Earth round the Sun, the apparent solar diameter varies from 31′ 31″ at aphelion to 32′ 35″ at perihelion, and the energy of sunshine rises correspondingly by 7 p.c. at the latter point. For reasons which will become clearer later on this cannot have much effect on the temperature of the surface, but in view of the great length of the lunar day, which averages 29 days 12 hours 44 minutes and 2·9 seconds of our reckoning, the additional heat may build up in the 'subsoil' and affect its temperature. As already explained in Chapter 3, the length of the lunar day will vary by reason of the joint orbital movement with the Earth, librations and axial inclination, but once more the consequences of these fluctuations are not climatically significant.

The only true seasons of the Moon are the times of its 'day' (p. 202). Since the Sun creeps very slowly across the lunar sky, the thermal situation on the surface depends substantially on the

Sun's altitude above the horizon alone, and the temperature declines more or less uniformly from the subsolar point towards the terminator. This means that the average midday temperature at, say, ±45° of latitude will be the same as at the *equator of illumination,* approximately 3 days and 18 hours after sunrise or before sunset, when the Sun is 45° up.

The situation is illustrated in Fig. 20, which shows the isotherms obtained by W. M. Sinton[27] from infrared readings with a Golay cell on a waxing gibbous Moon. The steady decline in temperature away from the subsolar point is clearly visible. The lowest temperature recorded near the terminator is −60° C, while the hottest areas exceed +110° C but fall short of +120° C. One of these areas includes the subsolar point, which lies in Mare Foecunditatis, but there is another area of high temperature in Mare Crisium to the north of it. The temperatures are averaged over some tens of square kilometres, and may be locally higher and lower. In fact, when the subsolar temperature is measured at the limb it comes out some 50° C lower than at the centre of the disk. This is explained by the preponderance of mountain slopes under oblique illumination within the sight of the thermocouple in the first case, but a cold thin atmosphere supporting a light frosty haze (say, of dry ice) could produce a similar effect, and we have seen in Chapter 15 that there is some evidence for its existence. In any event, the subsolar point in Sinton's chart being fairly close to the terminator, the reading may fall short of the true average. Still higher temperatures undoubtedly occur.

On the other hand, the mean temperature within 20° of the poles and the terminator never rises above freezing point, while midnight temperatures are put at some −145° C, derived from the microwave data. In fact, at such low temperatures thermocouple readings are not very reliable, as the level of energy of the infrared radiation to which the instrument responds falls very low. Microwave radio gives better results, but microwaves can penetrate a thickness of rock that increases with the wavelength, so that the result no longer refers to the actual surface, which will be still colder.[20, 21]

The isotherms have a jagged appearance and there are two hot areas, all of which bears witness to harsh local differences of temperature. In the absence of the tempering action of a dense atmosphere, all heat traffic is by radiation and absorption, and a shadow out of sight of the Sun and mountains that reflect some of the solar and radiate some of their own heat may be well below freezing while the sunlit part of the surface a few yards away is

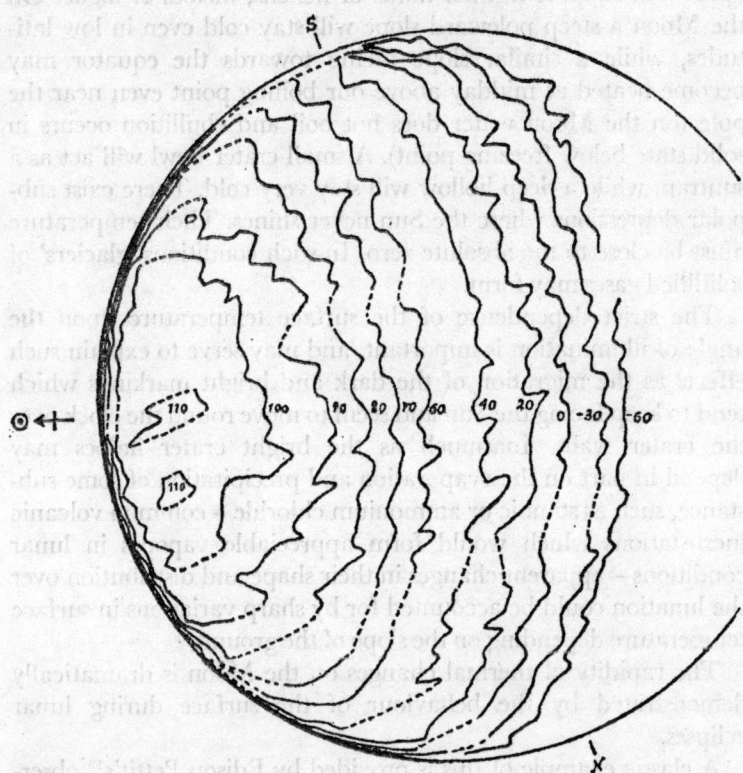

Fig. 20. Lunar temperature chart. Distribution of temperatures on a gibbous Moon after W. M. Sinton. The temperatures are given in °C. The subsolar point is in Mare Foecunditatis, but there is another hot area in the south of Mare Crisium below it. The arrow points toward the Sun. The jagged appearance of the isotherms reflects differences in the albedo and relief of the surface, untempered by convective circulation. It indicates the existence of microclimatic zones too limited in extent to be recorded by the detector.

above boiling point. Otherwise the distribution of temperatures is determined by the albedo and structure of the ground, as well as the orientation and angle of the slope. All these factors operate on the Earth too, but on the Moon they operate almost unchecked. Our botanists are familiar with the so-called micro-climates, where different sides of a boulder or a treebole may correspond to geographical climatic zones many degrees of latitude apart and support distinct floras of lichens, mosses or algae. On the Moon a steep poleward slope will stay cold even in low latitudes, while a similar slope facing towards the equator may become heated at midday above our boiling point even near the pole (on the Moon water does not boil and ebullition occurs in solid state below freezing point). A small crater bowl will act as a suntrap, while a deep hollow will stay very cold. There exist subpolar depressions where the Sun never shines. Their temperature must be close to the absolute zero. In such conditions 'glaciers' of solidified gases may form.

The strict dependence of the surface temperature upon the angle of illumination is important, and may serve to explain such effects as the migration of the dark and bright markings which tend to keep facing the Sun and seem to move round the clock over the crater walls. Inasmuch as the bright crater haloes may depend in part on the evaporation and precipitation of some substance, such as stannic or ammonium chloride – common volcanic incrustations which would form appreciable vapours in lunar conditions – apparent changes in their shape and distribution over the lunation could be accounted for by sharp variations in surface temperature depending on the slope of the ground.[6]

The rapidity of thermal changes on the Moon is dramatically demonstrated by the behaviour of the surface during lunar eclipses.

A classic example of this is provided by Edison Pettit's[19] observation of a small area near the centre of the disk during the total eclipse of 28th October 1939. At 4 hours 10 minutes G.M.T. the reading was 370° K ($+97°$ C). Fifteen minutes later the area entered the penumbra of the Earth's shadow and its temperature fell instantly by 10° C. Thereafter it grew colder and colder faster and faster, the total drop between $4^h 31^m$ and $5^h 30^m$, when the area was covered by the umbra, being from 357° to 198° K. The subsequent decrease of temperature was only gradual, indicating conduction of heat from the deeper surface layers. The lowest point reached was 177° K, or 57° C above the infrared night temperature of the surface (120° K). The totality ended at $7^h 45^m$, and by $9^h 14^m$ the temperature returned to and slightly exceeded the original figure of 370° K.

This is typical of the lunar surface as a whole, but considerable local anomalies exist. At the total eclipse of 19th December 1964 J. M. Saari and R. W. Shorthill[28] in the U.S.A. discovered hundreds of 'hot spots' on the Moon, mostly rayed craters, but also extended areas in the maria, and Mare Humorum in particular, which stay warmer than their surroundings and show bright in an infrared image of the eclipsed Moon. Tycho, Copernicus and Aristarchus have eclipse temperatures 40 – 50° C above the average. These results were confirmed and supplemented by G. R. Hunt, J. W. Salisbury and R. K. Vincent[10] during the eclipse of 13th April 1968. Scans taken 14 hours after the local sunset in Tycho show a heat-release pattern that clearly matches evening illumination and so points to the escape of heat stored up during the day rather than from a volcanic source. On the other hand, the thermal anomaly on the western shore of Mare Humorum cannot be explained in this way (p. 172).

Systematic differences in the rates of heating of the east and west limbs of the Moon after an eclipse were found on 7th July 1963 by M. N. Markov and V. L. Hohlova[16] in the U.S.S.R. This indicates differences in the subsurface temperature gradients, possibly associated with the formation of Mare Orientale and the subsequent, selenologically recent history of the western limb region. It also finds a counterpart in the different behaviour of crater rays and other bright markings near the two limbs (p. 205), strengthening the suspicion that this may be due to thermal causes.

Such local differences impose a limit on useful generalization. Diversifying forces have been at work, often on a large scale, it seems quite recently, and we have seen earlier on in this book that these forces could not have been purely external. The Moon is not simply sheathed in a uniform layer of dust or rubble, due to the meteoritic churning of some primordial unaltered material, as E. J. Öpik[18] maintains (1969). Yet, local variations apart, the salient fact that emerges from the eclipse observations is the extremely low thermal conductivity of the lunar surface and sub-surface materials. They are almost perfect insulators (*See* Appendix, pp. 255 f.).

A surface 'thermal skin', at most a few centimetres deep, heats up and cools very rapidly, but the heat of the Sun travels very slowly in lunar rocks and does not reach deep underground. At a modest depth of a metre or so the temperature will never rise above freezing point, although the surface above may be boiling hot. The 'subsoil' of the Moon must be in the condition of permafrost (p. 7).

This expectation is fully confirmed by radio observations in the millimetre and centimetre range (microwaves). The wavelengths of maximum energy radiated by a black body is by Wien's Law (*see* Appendix, p. 255) inversely proportional to its absolute temperature. The albedo of the Moon increases with wavelength,[25] so that it is not quite the theoretical black body, but this does not affect the general proposition that the energy maximum in the emission diagram moves to the right – in the direction of longer wavelengths – as the temperature continues to drop. At very low temperatures it shifts from the infrared region into the microwave radio register, where radio telescopes with their large apertures can be used to advantage (the received energy is proportional to the square of the aperture). Moreover, as already mentioned (p. 230), microwaves can pass through a thickness of solid rock material that increases with wavelength. This means that the longer the wavelength of observation is the deeper below the ground the main source of planetary radiation will lie, so that successive subsurface layers can be probed in this way.

The millimetre results do not differ appreciably from those obtained in the infrared, though the diurnal amplitude of variation is somewhat depressed.[14] When, however, the Australian radio astronomers J. H. Piddington and H. C. Minnett[20] examined in 1949 the lunar emission at the wavelength of 1·25 cm, whose main source they put at about 40 cm below the surface, the picture was significantly different.

In the equatorial belt they found a mean temperature of 249° K, rising to a maximum of 301° K and dropping to a night-time minimum of 197° K. The mean for the whole disk was 10° C lower, the diurnal fluctuation being 40·3° C on both sides of this figure. Moreover, the temperature maxima and minima lagged about 45° behind the position of the Sun in the lunar sky (solar colongitude, p. 202), and so $3\frac{1}{2}$ terrestrial days behind the local moon and midnight respectively.

These observations have been confirmed since and extended to other wavelengths. At wavelengths of over 20 cm the radio temperature remains substantially the same throughout the lunation, declining steadily as the wavelength and so the depth of origin of the radiation increases. The results are exemplified by the following figures : 21 cm, $250 \pm 30°$ K (Mezzer and Strassl, 1959); 21·6 cm, 245° K (Westerhout); 33·3 cm, 208° K (Denisse and Le Roux, 1957); 75 cm, $185 \pm 20°$ K (Seegert, Westerhout and Conway, 1957).[2] It will be seen that all these temperatures, except possibly the first, correspond to permafrost conditions. In 1964, however, the scientists of the Gorky Radio-Physical Research Institute in Russia announced the existence of a thermal discontinuity at about 30 metres below the surface, where the temperature was roughly 273° K (0° C).[7] The discontinuity would thus mark the lower boundary of the permafrost zone.

The Russians appeared to associate this boundary with a transition from the porous to solid state of lunar rocks, but this smacks too much of the common but improbable assumption that the Moon was formed by a single act and not by a continuous process of gradual changes : the fact that the ice in rock cavities could not extend beyond the point where the temperature rose above 0° C seems in itself sufficient to account for the thermal discontinuity of this kind.

Something like that must indeed be expected if the liberation of radioactive heat in the lunar crust is taken into consideration. Sooner or later a point will be reached where the initially declining temperature begins to rise again, and this will be sooner rather than later by the very fact of the low thermal conductivity of the crustal materials, which will be equally effective in sealing off the escape of heat from the interior as they are in shielding it from the heat of the Sun.

The nature of the materials having the thermal properties of the lunar surface has been hotly debated and cannot be determined at the present date.[8]

An early (1947) rigorous analysis of the eclipse data has led the Dutch investigator A. J. Wesselink[26] to conclude that even pumice is too good a conductor of heat to satisfy the requirements, which are best met by fine dust *in vacuo*. This lunar dust was inferred from polarimetric analysis by Lyot and Dollfus,[27] and has been greatly publicized in connection with Gold's hypothesis of the origin of the maria (p. 173). Not all of it seems to have settled yet.

One difficulty has always been that fine dust is a good reflector of light,[4] whereas the mean albedo of the Moon is only 0·07. It, therefore, had to be 'dark dust'. Yet in comparing the moonlight at successive phases with the light reflected at corresponding angles of illumination by various terrestrial materials Jan van Diggelen of Utrecht (1960) found that the false reindeer moss *Cladonia rangifernia* gave the best fit. From this he concluded that 'The Moon's surface is not merely a layer of dust ..., but has an irregular spongy character.'[24] The close-ups of Oceanus Procellarum obtained by the Soviet 'soft-lander' *Luna IX* (Plate 19a), are in close agreement with this result. There is no unconsolidated dusty overlay, and the material appears to be stiff and brittle withal. The *Surveyor I*, on the other hand, gives a somewhat different impression of a nearby area of the Oceanus, where a thin overlay having an initially soft dusty texture (volcanic ash?) is present (p. 175 and Plate 19b).

The trenching operations carried out by the *Surveyors* reveal a soil-like granular aggregate, where particles varying from the millimetre to centimetre size and occasional larger rock fragments are loosely cemented to give an overall bearing strength of 'wet sand', estimated at 4 to 7×10^5 dynes/cm^2.[22] This could perhaps be assimilated to Lyot's description of a reflecting layer of opaque material 'ground to very fine grains which themselves are combined into larger grains',[27] but no loose dust has been found. The blast of the vernier rockets at the touch-down more particularly of the *Surveyor III,* where they failed to shut off on landing, so that the probe made several hops before coming to rest, had left no detectable impression on the lunar surface, and the special 'blowing experiment' made by the *Surveyor I* gave a similarly negative result.

It may be that lunar dust is cemented by the sputtering due to the impact of solar protons and cosmic rays, as well as the partial vaporization and deposition of material in micrometeoritic bombardment, giving it a semi-rigid 'fairy-castle' structure. If so, however, it can no longer be properly described as 'dust' and Wesselink's conclusion applied to it, as is done by Öpik (1969).[18] In any event Wesselink overlooked the point that the conductivity of granular material depends on the pressure at contact between the grains (Firsoff, 1959),[6] and this in turn is a function of gravity, so that the laboratory results obtained on Earth cannot be applied to the Moon without modification. He himself appears to have accepted this criticism (private communication), but not so those who merely refer to him.

With more recent data Harper and Jaeger (1950)[11] found that the thermal behaviour of the Moon's surface 'skin' can be matched by terrestrial pumice (specific gravity $0·6$ g/cm^3) overlain by a layer of airless dust 1 millimetre thick.[12] In J. H. Fremlin's[8] analysis a patchy dust cover, leaving $4·8$ p.c. of bare solid rock clear, would do (1959). The Russian investigators Matveyev, Suchkin and Troitsky (1965)[17] have inferred from a comparison of passive-radio and radar data on the reflection coefficient and its variation with wavelength that the thermal behaviour of the lunar surface can be explained by an increase in density by a factor of about $1·5$ over the first 4 cm, whereafter the density remains constant. The conductivity of a material so structured would lie somewhere between that of 'small dispersed dry substance and solid porous state of terrestrial rocks'. Laboratory reticules (p. 58) can easily meet these requirements, but the possibility of massive, if transitory, water action in the maria cannot be ruled out. Their material may, after all, be at least partly mud congealed under low gravity. The position remains somewhat obscure.

In a report on the *Surveyor V* (1968) Christensen *et al.*[1] write : 'Best agreement obtained for a compressible soil model with the observer *Surveyor V* foot pad penetration and landing leg loads is for a soil static-bearing capability of $2·7$ newtons/cm^2* and a density of $1·1$ g/cm^3. Incompressible soil model analyses have not yet been performed for *Surveyor V*'.

I am always suspicious of models, and in this case the proposed model must meet not only the mechanical, but the thermal requirements as well.

The mechanical grab of the *Surveyor VII*, which landed on the outer slopes of Tycho, picked up a rock fragment. This was weighed and its volume estimated from television views at different angles – no doubt, a wonderful performance that! The resulting density came out at between $2·4$ and $3·1$ g/cm^3. The surrounding soil, in which this and other stones were embedded, is assumed to have been formed by the fragmentation of rock of

* Newton = 1 kg.m.s.$^{-2}$

the same density by meteoritic bombardment. Hence it is opined (R. F. Scott and E. L. Roberson, 1968) that 'This would appear to reinforce the conclusions obtained from the *Surveyor III* surface-sampler operations that, in fact, the strength and deformation characteristics of the lunar-surface granular material can be explained by the presence of a material with a density comparable to that of common terrestrial soils, that is in the range of 1·5 g/cm^3 and greater'.[22] No dust for this once. . . . . .

These results may fit very nicely H. C. Urey's idea of a homogeneous Moon, but how are they supposed to fit the thermal behaviour of the lunar surface at eclipses matching that of fine dust *in vacuo*? (For Wesselink's analysis, if not his conclusion, stands confirmed).[13] This does not seem possible. We may soon have the complete answer. Meanwhile let us return to that rock sample. Its density would correspond to that of our abyssal igneous rocks, which are not formed on the surface, but hundreds and thousands of metres beneath it under the pressure of heavy overlying strata acted upon by a gravity six times the lunar. Long ages of intense erosion are needed to expose these rocks on the surface of the Earth. It is hard to conceive of a geological process that could account for the presence of such a rock on the surface of the Moon. On the other hand, there is every reason to expect finding meteoritic rocks there, and the *Surveyor VII* specimen may have been a stony meteorite. Thus in the one case where the meteoritic hypothesis could have been profitably invoked it is completely overlooked. At the same time we must not forget that stony meteorites do not arise naturally in airless space under zero gravity; their genesis involves high temperatures and pressures of formation. Such rocks could, therefore, no more be primary constituents of the lunar surface than granite or gabbro. Conjecture apart, the high-resolution *Orbiter* views of the Alps and other lunar mountains show a structure that is nothing like the weathering patterns of abyssal igneous rocks. It is characterized by concentric flow lines, often pivoted on small craters and suggestive of a slow movement of relatively cold viscous lava. There are also peculiar cellular grids, for instance, in the crater Fauth, which could be explained by the weathering of friable porous rock with indurated joints.

On whatever reckoning, porous or fragmented materials, whether consolidated into fairy-castle texture or not, must be present on the Moon, and all of them are active gas sorbents. I have considered the problem of sorption in *Strange World of the Moon*[6] in some detail and will not go over it all again here. The important point is that *sorption,* which is a blanket term for physical adsorption and absorption by solution, as well as the absorption by chemical bond, although sensitive to the changes of temperature, is little affected by the partial pressure of the sorbed gas or the total barometric pressure.

Sorption can be represented by an empirical formula $x/m = k\,p^{1/n}$, where x stands for the mass of the sorbed gas, m for that of the sorbent and k is a constant peculiar to any pair of substances involved in the process. The value of n varies from 1 to 10, being usually about 5, and is, moreover, inversely affected by falling temperature.[15]

To quote from MacBain's *The Sorption of Gases and Vapours by Solids* (p. 34 f.) :

> The fact that the sorption curve is continued down to the lowest pressures is of great practical significance. This is forcibly illustrated by considering the effect of diminished pressure upon the sorption of vapour for which $n = 5$; that is, the sorption varies as the fifth root of the pressure. To halve any given value of sorption, the pressure would have to be diminished 32-fold. To reduce a given value of x/m to $0\cdot1$ p.c. of its previous value, the pressure would have to be diminished in the proportion of 10^{-15}-fold, an operation which cannot be accomplished experimentally . . .

Now if we put the ground-level barometric pressure on the Moon, following Elsmore, at 10^{-13} of an atmosphere, this would still fall two orders of magnitude short of the operation envisaged by MacBain. Nor must we forget the very low night temperatures, of about $-150°$ C or less, which will raise the value of n. Thus even on this extremely unfavourable assumption appreciable gas sorption must be expected in lunar conditions.

Adsorption is almost instantaneous, and at low temperatures multiple molecular layers of adsorbed gas will form on the surface of the pores and grains. Absorption by solution and chemical bond (hydration, ammoniation, etc.), as well as condensation may occur. Reactions between the gases and between them and the solids may be expected, as good sorbents are also effective chemical catalysts. Silicates have a special affinity for water vapour, which also forms about 90 p.c. of volcanic exhalations, so that most of the sorbed gas may be water vapour, with carbon dioxide coming next. The gas will percolate through the interstices of the ground to the permafrost layer. As the temperature of the sorbent layer fluctuates over the cycle of the lunar day, some of the gas will be released into the atmosphere, but the gases within the permafrost layer and especially that part thereof where the temperature stays the same day and night will be permanently held. In localities particularly favourable to sorption a situation which I have described as *gas marsh*[6] may develop, and capillary condensation of water is not beyond the bounds of possibility. This situation could give a foothold to simple organisms, and even plants are conceivable which found all their sustenance underground and only emerged above it for the energy of sunlight. As far back as 1900 Macfadyen showed that various organisms could survive immersion in liquid air ($-193°$ C) for 20 hours, and five years later Becquerel demonstrated that not only seeds, bacterial spores, algae, lichens and mosses, but such animalculae as rotifers and tardigrades can withstand a temperature of $0.0075°$ K and exposure to hard vacuum.[28] These tests are far harder than even a lunar environment can impose, and the organisms subjected to them have evolved on Earth where they had no need to adapt themselves to such conditions, so that, although speculative, these reflections are by no means far-fetched.

Be this as it may, sorption also provides a basis for a permanent or semi-permanent daytime atmosphere close to the surface in addition to the tenuous gas halo disclosed by the occultations of radio sources. We have seen in Chapter 15 that there are many indications of the actual existence of such a daytime atmosphere, although even the issue of a more conventional gas envelope cannot be considered closed.

As already stated, some of the gas held by sorption in the lunar surface will be released in the daytime heat. The initial release may be rapid in concordance with the thermal behaviour of the surface 'skin' during lunar eclipses. But the latent heats of sorption are quite high, between 3,000 and 12,000 small calories per mole of the sorbed gas of vapour.[15] Surface and subsurface precipitates may likewise be present. The vapours concentrated by sorption in the interstices of the ground may freeze as the temperature continues to drop to the midnight minimum. To revaporize these precipitates, the latent heat of vaporization must be supplied as well. The latter is comparable to the latent heat of sorption, but unlike it, it depends on the nature of the substance and the barometric pressure only, there being no interaction with the sorbent itself, which affects the heat of sorption. This alone may suffice considerably to delay the downward progress of the solar heat and give a false value for the conductivity of lunar rocks as calculated on the assumption that everything happens in a vacuum, which may serve further to illustrate the unwisdom of uncritically applying laboratory findings to the complex field conditions. In any event the progress of degassing will be greatly slowed down as it reaches the deeper subsurface levels, while the volume of the liberated gas or vapour may increase at the same time.

It is also clear that the exhaled vapours will be very cold, and may even become recondensed upon emergence as they are chilled in adiabatic expansion. Under very low atmospheric pressure and at relatively high daytime temperatures obtaining on the Moon some salts, among which stannic and ammonium chlorides have already been mentioned, and other substances, such as sulphur, may form appreciable vapours, contributing heavy components to the daytime atmosphere, which may vary considerably in composition from place to place. In being reprecipitated at some stage or other of the degassing process, they may form temporary 'frostings' on the surface, though at low temperatures, early in the morning or late in the evening and near the poles, ordinary hoarfrost or dry ice (frozen CO_2) may appear as well. Thus the lunar surface would be fuming gently, perhaps imperceptibly, although a frosty haze of fine crystals may form at the top of the evolving atmosphere, as some observations, including my own ones of star occultations in blue (p. 221), and the *Orbiter* frame V–52M (Plate 20A), appear to indicate.

On any reckoning the desorbed vapours, being cold and heavy, will rise only sluggishly above the surface, in which they should be further hindered by electrostatic forces. Desorption is accompanied by ionization, the escaping gas molecule leaving behind an electron as it breaks free of the sorbent. Thus the desorbed gas acquires a positive charge and the sorbent a corresponding negative charge. The evolving gas will, therefore, tend to be electro-statically bound to the surface. In these circumstances glow discharges are likely and may account for some of the reported lunar glows (pp. 188 f.).

Whether the general overlying atmosphere is directly compar-able to our exosphere or not depends on the accuracy of Elsmore's oft-quoted figure, but there can be no doubt that it is very thin, and the molecules at the top of the newly desorbed gas layer will diffuse into it. As already mentioned on p. 216, such molecular evaporation will remove the most energetic molecules from the gas – and will remove them quite fast, as a result of which the temperature of that top layer will tend to drop towards the con-dition where further dissipation ceases. It may never attain this condition, but it will be very effectively chilled all the same. Chill-ing entails volume contraction, so that as though a skin of increased density not unlike the tension skin on a liquid, will form at the top of the dissipating gas. Further ionization in the short-wave radiation of the Sun will have a similar effect, as the free electrons will escape very rapidly, leaving behind a growing positive charge, and this in turn will induce a negative charge in the subnatant gas, the latter striving to pull the positively charged gas down towards it.[5] In other words, the daytime atmosphere will have a well defined upper boundary, where condensation may occur.

The final effect must depend on the amount of gas held captive in the surface formations and the proportion of it released during the day. In the absence of any numerical data it is wellnigh impossible to attempt an estimate, but we have seen that even with a ground particle density of 10^6 sorption will operate quite effec-tively. We have also seen that gases continue to be emitted from the interior and some volcanic activity may be in progress. This may help to raise the overall density of the daytime atmosphere, possibly to as high as 0.0001 to 0.001 of our sea-level value. Since this atmosphere tends to keep close to the ground, local gas pools of higher density may develop, for instance, within volcanically active craters and mountain rings, such as Aristarchus or Tycho, whose thermally abnormal behaviour (p. 233) may be related to a shielding layer of gas rather than to differences of surface struc-ture.

The envisaged atmospheric regime is cyclic. Indeed, it is the alternation of long cold nights and equally long hot days that makes it feasible. At nightfall the lowermost atmospheric layer will be absorbed almost at once and replaced by the overlying gas, which will be depleted in its turn as sorption progresses and accelerates with decreasing temperature. Thus the atmosphere is undermined in its densest part and is gradually pulled down and soaked in, until only the thin outer halo is left.

The recurrent sorption and desorption also provides a powerful mechanism of erosion. Intensitive *persorption* (adsorption in depth) leads to the dilatation of the molecular lattices of the sorbent, which may as a result be reduced to fine powder, as were metals in Gurwitsch's experiments.[15] Porous rocks are virtually immune to exfoliation, which is the most usual form of terrestrial heat erosion (it requires the presence of some moisture withal – Blackwelder), but they will be preferentially attacked by sorption by virtue of being better sorbents than the more compact formations. This will lead to the eventual isolation of the more compact rock masses as boulders or tors. On the other hand, there is no reason to expect such rocks to be spared by micrometeoritic, protonic or cosmic-ray bombardment. It is, therefore, significant that large boulders and summit tors, closely comparable to the tors of Dartmoor, exist on the Moon. Several such will be seen in the *Orbiter* photographs III–199H$_3$, and III–184H$_3$ (Plate 8b), which show portions of the Flamsteed-P ring, and whole crowds of them in V–151H$_1$ on the central mountain group of Copernicus. These rocks are no ejecta, which would accumulate in hollows and at the foot of steep slopes, as they tend to occur on the summits, steep and sometimes narrow ridges, edges of mountain scarps and crater lips. A few loose boulders which have rolled down the slopes can be easily identified and present no problem. This stamps them clearly as products of erosion, which must have removed many metres of material before the larger tors could have been isolated. Yet the Alps, the Apennines and other lunar mountains are totally devoid of screes.

A typical scree talus emanates from a gully and has a conical surface, which, seen from the side, is concave at the tail, where most of the heavier boulders are to be found. With all due respect to J. E. Guest and J. B. Murray[9] this is nothing like the structure of the ground at the foot of Tsiolkovsky's inner glacis. In fact, this ground is closely comparable to the vast hinterland of the Alps and the broad forelands of the Orientale mountain ranges : whatever it may be, it is not scree. Nor do the 'welts' fringing some heights, say in Flamsteed P (II–199H₃), with their terminal swellings and steep fronts, bear any resemblance to scree slopes: they are explained much more naturally by the arrested flow of viscous lava (p. 134). Incipient screes may be in the process of developing below the Copernicus peaks, but they have not progressed very far yet. Evidently in lunar conditions, where ground water and strong wind are absent, boulders do not roll down as readily as on the Earth.

On the other hand, there is no mistaking the downward flow of loose material on the smooth and steep inner glacis of some craters, well exemplified by Mösting in frame III–II3H₂. Whether, though, the origin of this material is erosive is highly doubtful. It seems far more likely to have been ejected from the volcanically active rim fracture, along which the crater wall has grown. As a rule the fracture is sealed and can only be inferred, but frame V–70H₂ gives a very instructive high-resolution view of Dawes, whose inner glacis are very steep and uniform, and the bounding ridge unusually sharp, the rim fracture still gaping open along part of it.

Sorption, however, may operate as a binding agency as well. As previously pointed out, good sorbents are also chemical catalysts, and so may mediate reactions between loose grains and sorbed gas, binding these together in a kind of loose pseudo-sinter (true sintering involves heat-welding at contact). Thus sorption could be responsible for the consolidation of lunar soils.

Some of the questions raised in this chapter could be answered by the examination of the samples of lunar ground that will be available, following the *Apollo XI* voyage to the Moon. It will be interesting to compare the suggestions made here with the outcome of this examination. Other problems, however, cannot be solved as easily as that and may require years of sustained on-the-spot investigation.

REFERENCES

1 CHRISTENSEN et al. (1968). *Journal of Geophysical Research*, **12**, p. 7,155 ff.
2 FIELDER, GILBERT (1961). *Structure of the Moon's Surface*. Pergamon Press, Oxford.

3   FIELDER, G., and GUEST, J. E. (1968). 'Size-frequency Distribution of Particles and Lunar Surface Materials'. *Earth and Planetary Science Letters 5*, p. 86 ff.

4   FIRSOFF, V. A. (1959). 'Is the Moon Covered with Dust?', *J.B.A.A.*, **69** (4), p. 153 ff.

5   FIRSOFF, V. A. (1959). *Science*, **130**, p. 1,337.

6   FIRSOFF, V. A. (1959). *Strange World of the Moon*. Hutchinson, London.

7   FIRSOFF, V. A. (1964). 'The Surface of the Moon', *Discovery*, **25** (6), p. 25 ff.

8   FREMLIN, J. H. (1959). 'Sub-surface Temperatures on the Moon', *Nature*, **183**, p. 1,317 f.

9   GUEST, J. E., and MURRAY, J. B. (1969). 'Nature and Origin of Tsiolkovsky Crater, Lunar Farside', *Planet. Space Sci.*, **17**, p. 121 ff.

10 HUNT, G. R. *et al.* (1968). 'Infrared Images of the Eclipsed Moon', *Sky and Telescope*, **36** (4), p. 223 ff.

11 JAEGER, J. C., and HARPER, A. F. A. (1950). *Nature*, **166,** p. 1,026.

12 JAEGER, J. C. (1959). 'Sub-surface Temperatures on the Moon', *Nature*, **183**, p. 1,316 f.

13 LOW, F. J. (1965). 'Lunar Night Temperatures Measured at 20 Microns', *Ap. J.*, **142** (2), p. 806 f.

14 LOW, F. J., and DAVIDSON, A. W. (1965). 'Lunar Observations at a Wavelength of 1 Millimetre', *Ap. J.*, **142** (3), p. 1,278 ff.

15 MACBAIN, J. W. (1932). *The Sorption of Gases and Vapours by Solids*. Routledge, London.

16 MARKOV, M. N., and HOHLOVA, V. L. (1965). *Astronomičeskij Žurnal*, **42** (2), p. 386 ff.

17 MATVEYEV, Y. G., SUCHKIN, G. L., and TROITSKY, V. S. (1965). *Astronomičeskij Žurnal*, **42** (4), p. 810 ff.

18 ÖPIK, E. J. (1969). 'The Lunar Environment', *Science Journal*, **5** (5), p. 67 ff.

19 PETTIT, EDISON (1940). 'Radiation Measurements on the Eclipsed Moon', *Ap. J.*, **91**, p. 408 ff.

20 PIDDINGTON, J. H., and MINNETT, H. C. (1949). 'Microwave Radiation from the Moon', *Australian Journal of Scientific Research*, Series A, **2** (1).

21 PIDDINGTON, J. H. (1961). *Radio Astronomy*. Hutchinson, London.

22 SCOTT, R. F., and ROBERSON, F. I. (1968). 'Surveyor VII, Mission Report, Part II', *NASA – Technical Report 32–1264*.

23 SHORTHILL, R. W., and SAARI, J. M. (1966). 'Recent Discovery of Hot Spots on the Lunar Surface ...' in *The Nature of the Lunar Surface* – Proceedings of the 1965 IAU–NASA Symposium. Johns Hopkins, Baltimore, Md.

24 STRUVE, OTTO (1960). 'Photometry of the Moon', *Sky and Telescope*, **20** (2), p. 70 ff.

25 WATTSON, R. B., and DANIELSON, R. E. (1965). *Ap. J.*, **142** (1), p. 16 ff.

26 WESSELINK, A. J. (1948). 'Heat Conductivity and Nature of the

Lunar Surface Material', *Bulletin of the Astronomical Institutes of the Netherlands,* **10** (390), p. 35 ff.

27 KUIPER, G. P., and MIDDLEHURST, B. (Eds.) (1961). *Planets and Satellites. The Solar System,* Vol. III. University of Chicago Press.

28 MAMIKUNIAN, G., and BRIGGS, M. (Eds.) (1965). *Current Aspects of Exobiology.* Pergamon Press, Oxford.

Abbreviations used

Ap. J.	=Astrophysical Journal.
J.B.A.A.	=Journal of the British Astronomical Association.
Planet. Space Sci.	=Planetary and Space Science.

Appendix

This part of the book gives a technical and mathematical elaboration of some selected issues from the main text with reference to chapter or page. The subject is a vast one and full treatment of the standard problems will be found in appropriate textbooks, which there would be little point in duplicating, so that the basis of selection must necessarily be narrow, only such matters being given consideration as are in any way unfamiliar, novel or especially pertinent to the views proposed in this work.

P. 16

A solid substance below its melting point may yet, if confined under a sufficiently high pressure, deform continuously without fracture, and is then spoken of as *rheid*. Such 'solid flow', the best example of which is that of glacier ice, is measured by *rheidity* (S. Warren Carey), defined as a ratio of viscosity to rigidity, which has the dimension of time, and, being a small number, is usually multiplied by 1000 when *c.g.s.* units are used[8]:

$$\text{Rheidity} = 1000 \, \frac{\text{viscosity}}{\text{rigidity}}. \tag{1}$$

For instance,

$$\text{Rheidity of ice} = 1000 \times 10^{13} \, / \, \frac{}{10} = 10^6 \text{ seconds.}$$

P. 58

Terrestrial lavas are often highly vesicular and have bulk densities of about $\frac{1}{4}$ of the same rock material in compact state. If V' be the volume of gas bubbles in a lunar lava at a depth h and a given temperature and V'' the corresponding volume of bubbles in the same lava on the Earth, Fielder[4] gives the following ratio between the two

$$\frac{V'}{V''} \simeq 6(1 + P/\rho\, gh), \tag{2}$$

where P is the atmospheric pressure at sea level, ρ the bulk density of the overlying materials, and g terrestrial gravity. We may call this ratio ψ. The following table has been obtained on this basis:

h	1 km	100 m	10 m	1 m	10 cm
ψ	6.0	6.3	8.6	30	260 .

Since, however, the bulk density of the overlying lunar materials must also be lower than that of their terrestrial counterparts, these figures may be taken to represent the lower limit.

P. 135

In considering the flow of lava on the lunar surface O'Keefe *et al.*[11] make use of an equation proposed by Jeffreys and modified by Nichols:

$$V = (g \sin A \, d^2\rho)/3\mu , \tag{3}$$

where V stands for the velocity of flow, g for gravity, A for the angle of slope, d for the depth of the flowing layer, ρ for its density and μ for its viscosity. Putting $d = 300$ m, $A = 16°$, $\rho = 3$ g/cm^3 and $\mu = 10^5$ in c.g.s. units as the upper limit for the viscosity of basaltic lavas, they obtain $V = 1$ km/sec $= 3600$ km/hour. This result is clearly absurd: no basaltic lava or any other liquid could flow at such a rate on the Moon or anywhere else. A lava would either be immediately transformed into a fluidized ash flow or greatly expanded by vesiculation (cf. the foregoing section), which must involve surface cooling through the expansion of the evolving gas bubbles, and would in any case reduce its specific gravity and increase its viscosity. The equation is inapplicable in either case.

P. 211

Jeans's theory of dissipation of planetary atmospheres will be found on p. 343 ff. of *The Dynamic Theory of Gases*[10] and has been summarized *inter alia* in the Appendix to my own *Strange World of the Moon*[6] (p. 193 ff.).

He assumes that the atmosphere is in isothermal equilibrium and rotates solidly with the planet. On these assumptions it may, then, be said to extend at the equator up to a distance where gravity and the centrifugal force of rotation draw equal. At this theoretical limit the atmosphere would behave as though it were all streaming away with a uniform velocity $\frac{1}{4}C$, where C is the mean square-root molecular velocity, such that $C^2 = \dfrac{3RT}{m}$, R

being the Gas Constant $= 8.315 \times 10^7$ ergs per mole, T the absolute temperature of the gas (and so of the isothermal atmosphere), and m the molecular weight. This situation will be exactly parallelled in an actual atmosphere so constituted at a theoretical interface above which molecular collisions become so infrequent that they can be neglected if we substitute for C the mean excess

of molecular velocities over the velocity of escape $U = \sqrt{\dfrac{2ga^2}{R}}$,

where g is surface gravity, a the planet's radius and R the distance of the escape level from the centre of the planet. This can further be represented as the escape of the same amount of gas from an atmosphere of uniform density and temperature and having a mass equal to that of the actual atmosphere through a series of orifices that allow the required amount of gas to pass through. To this extent it is immaterial at what level the escape occurs.

Since C^2 is inversely proportional to m, in a gas mixture any constituent may be treated as a separate atmosphere escaping at a rate corresponding to its molecular, or particle, weight. On these assumptions Jeans derives a formula for the time t_1 in which the isothermal atmosphere of a given gas, characterized by C, will be totally lost by the planet:

$$t_1 = \frac{C^3}{2g^2 a} \cdot \exp\left(\frac{3ga}{C^2}\right). \tag{4}$$

This treatment gives us a measure of understanding of the process, but, apart from the mathematical error that t_1 does not give us the time of total loss but only the time in which the mass of the

atmosphere will drop to $\dfrac{1}{e}$ of its original value[15] (e being the base

of natural logarithms and ~ 2.72), the initial assumptions are variously unrealistic.

Planetary atmospheres are not isothermal. It is a fallacy to assume that all atmospheres must necessarily have the same structure as ours, but this may legitimately be regarded as a fair model for terrestrial planets where the range of temperatures and other conditions do not differ greatly from those of the Earth. In such an atmosphere the lower portion or troposphere is in convective equilibrium, which approximates the adiabatic condition where $pv\delta = $ const., p being the pressure and v the volume of a constant mass of gas, while $\gamma = c_p/c_v$, or the ratio of the specific heat of the gas at constant pressure to its specific heat at constant volume. (The value of γ depends on molecular structure and is approximately 1.67 for monatomic, 1.41 for diatomic and 1.26 for triatomic molecules.) In such an atmosphere $dT/dz = $ const. $= \Gamma^* = g/c_p$ ($T = $ absolute, temperature, $z = $ height above the ground, $g = $ surface gravity). This means that the temperature of the atmosphere declines uniformly with height, and so if we start with the same ground pressure p_0 and the same ground temperature T_0, the first of which is determined by the total atmospheric mass, at any given height z the adiabatic atmosphere must inevitably be denser (as well as colder) than the equivalent isothermal atmosphere.

There exists a limiting theoretical height Z at which the density of an adiabatic atmosphere drops to 0. This is

$$Z = \frac{\gamma C^2}{3g(\gamma-1)} \text{ (notation unchanged)}. \qquad (5)$$

But in actual fact this height is never reached, because viscosity is by Maxwell's Law largely independent from the density of the gas ρ, while the effectiveness of convection rests on the differences in specific weight of gas masses of different temperature: thus ascending currents lose their punch with increasing rarefaction and are halted at the tropopause. Above it the atmosphere assumes isothermal structure, in which it would continue except for the effects of photo-dissociation and the consequent formation of an ozonosphere, which in being heated by the absorption of the solar ultraviolet radiation creates a secondary convective zone, and the high temperature regions due to ionization. In the topmost reaches the atmosphere tends to become stratified according to atomic weight.

This is nothing like the situation envisaged by Jeans. The correction introduced by Lyman Spitzer, Jr.[15] is meant to bring the theory into better agreement with reality. By itself, however, it is not enough.

In applying any theory we must never forget the assumptions made at the start, on which its validity depends. This simple principle is greatly sinned against.

Firstly, Jeans's reasoning applies to a uniformly mixed atmosphere of ideal gases. It does not apply to vapours, which may be frozen out of the upper atmosphere at corresponding cold traps, and no molecule can escape to space unless it is bodily present at the escape level. Thus, for instance, the fact alone that water vapour of a molecular weight 18 may be too light to be retained by a given planetary or satellitic body does not necessarily mean that it will be lost to space in the way and at the rate indicated by the misapplied theory; it must reach the escape level first, which it may be prevented

*Adiabatic lapse.

from doing by the cold trap. Stratification at the top of the atmosphere may also shield heavier molecular species from escaping to space. On the other hand, photo-dissociation may expedite their loss.

Secondly—and this is fundamental!—Jeans assumes that the process of escape does not affect the temperature of the escape layer (layer rather than level, which is a fictional interface).

C is only a theoretical *mean* velocity, which corresponds to the condition of equipartition of energy, where every particle has the same kinetic energy $\frac{1}{2} mc^2$ (c stands for the actual molecular velocity and not that of light!). In reality, however, as the molecules collide and rebound, their velocities are subject to constant fluctuation, and if represented in a graph with velocities in abscissa and numbers in ordinate, the resulting curve would look something like this:

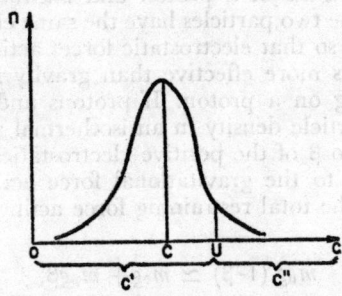

where the various values of c would have a maximum density at C and drop off regularly in the direction of lower and higher velocities. If U be the velocity of escape, we may call c' all those velocities where $c < U$ and c'' all those velocities where $c \geqslant U$. Obviously only the molecules in the class c'' can escape to space.

Consequently, the numbers in class c'' will be decreasing, while those in the class c' will increase in relative proportion. This implies a constant drain on the kinetic energy of the system, which will tend to readjust itself at a lower equipartition velocity C, until the numbers of molecules in the class c'' have become negligibly small, i.e. escape to space has effectively ceased.

This means that the mechanism of escape considered by Jeans can operate only if heat is supplied to the escape layer, to keep it at a constant temperature. No such source of heat exists, because the atmosphere at the start of the escape, e.g. when it comes into full sunlight through rotation, will attain a radiative equilibrium temperature, at which the solar input and the radiative output of heat by the atmosphere cancel out. Escape represents a net loss. To some extent this loss may be made good by the influx of more energetic molecules from below. Yet these molecules move by definition in the part of the atmosphere where molecular collisions are too frequent to be neglected, whereas those molecules that escape to space encounter no effective opposition. Therefore, the rate of the escape of the latter must be faster than that of the influx of the former: the loss of kinetic energy and so the decrease in temperature will continue.

The mechanism of molecular dissipation is self-defeating. This is only a particular example of the operation of the Le Chatelier principle whereby the products of any physical process oppose its continuation.

P. 217

The problem of escape of electrons from an atmosphere of ionized hydrogen has been considered by van de Hulst[26] (1954).

Let m_p be the mass of a proton and m_e that of an electron. $m_p/m_e \sim 1840$. The two particles have the same electrostatic charge e of opposite sign, so that electrostatic forces acting on an electron will be 1840 times more effective than gravity in relation to the same forces acting on a proton. If protons and electrons are to have the same particle density in an isothermal atmosphere under gravity g, the ratio β of the positive electrostatic force, which will repel the proton, to the gravitational force acting upon it must equal 0.5. Then the total restraining force acting on a proton will be

$$m_p g \ (1-\beta) \simeq m_e g + m_p g \beta, \tag{6}$$

the expression on the right being the restraining force acting upon an electron. The restraining forces being equal, both kinds of particle will escape at the same rate.

This condition, however, cannot be satisfied without a positive atmospheric charge[7], and this will increase with the weight of the positive ion m, to which it is proportional, as the electrostatic force repelling the positive ion is the same that attracts the electron and, the term $m_e g$ being negligibly small, the condition of parity is achieved with the electrostatic force $f = \frac{1}{2} mg$.

The situation, however, is more complicated than that, for the positive charge in the affected atmospheric layer will immediately induce a compensatory negative charge in the subnatant layer or the surface, whichever the case may be, and this will hold back the ions, preventing or retarding their escape, while it will repel the electrons. The latter will, therefore, continue to escape even after the theoretical condition of equilibrium defined above has been reached[5,7]. Indeed, van de Hulst himself anticipated this to some extent in remarking that 'In an isothermal gas the light electrons have a tendency to segregate themselves from the heavier protons'[26], although this may be interpreted as implying simple segregation according to particle weight in a gravitational field, where alone the isothermal condition would be relevant. In any event he was thinking of the solar and not a planetary atmosphere.

The mutual attraction between the positively-charged and the negatively-charged layers will tend to draw them together and increase their density. The chilling due to molecular evaporation to space, considered in the foregoing section, will have a similar effect of condensing the escape layer of gas and causing it to sink owing to its increased weight. Thus this layer would tend to develop a well-defined outer boundary and be comparable in some ways to the tension skin on liquid surfaces.

Chapters 11 and 15, and p. 225 f. in particular.

When a radio source is occulted by an atmosphered body the refraction of the radio waves depends on the charge density and its distribution with height above the surface of the body. If each particle carries a single electronic charge e and its weight is m the resonance frequency of the plasma f_0 observes the following relationship:

$$f^2_0 = \frac{N.e^2}{\pi m}, \tag{7}$$

where N is the number of particles per cm³; and the refractive index n for waves of the frequency f is given by the equation:

$$n^2 = 1 - f^2_0/f^2. \tag{8}$$

Since m occurs in the denominator of (7) and the mass of an electron is 1840 times smaller than that of the lightest possible ion, proton, electrons alone make an important contribution to the refractive index[14].

In calculating atmospheric densities from electronic refraction it is usually postulated that the number of electrons is the same as that of positive ions, which is fixed in proportion to that of neutral particles depending on the atmospheric model. Thus Elsmore[2] assumes that 1 in every 1000 molecules of the lunar atmosphere is ionized; with an electron density of 1000 per cm³ this gives a total density of 10^6 per cm³ or so. We have seen, however, on p. 251 that this is not a legitimate assumption. The whole procedure is highly speculative in any case. There are reasons to believe that the outer halo of the lunar atmosphere consists mainly of argon, possibly derived from the radioactive decay of potassium-40. Argon has a high ionization potential (15.68 electron-volts [N, 14.48; O, 13.55; H, 13.52]) and may not be ionized to the extent assumed by Elsmore, but if we take it that it is, then by the reasoning of the preceding section there will be at least 20 times as many argon ions as there are electrons, and Elsmore's figure of 5×10^{-13} becomes 10^{-11} of the terrestrial sea-level density at STP. The presence of still heavier ions, envisaged by Elsmore, will further increase this figure.

In his paper *Radio Observations of the Lunar Atmosphere*[2] Elsmore makes a surprising statement (p. 1044): 'Whereas the earth's atmosphere above 400 km consists almost entirely of N_2 and O_2, the atoms and molecules comprising the 'permanent' lunar atmosphere must have a molecular weight greater than 42 if the temperature is equal to, or greater than the surface temperature of 380° K; otherwise the molecular velocities will exceed the velocity of escape.' It would be hard to imagine so many inaccuracies finding their way into a single sentence by mere chance. To begin with, 380° K is not the temperature of the lunar surface, but only of its subsolar point. The mean grey-body temperature is 273° K (p. 255). The temperature of a rarefied atmosphere may be expected to be considerably below that of the surface, at least in its lower reaches. Yet even if we suppose that it is 400° K and accept Elsmore's improbable condition that the atmosphere is isothermal, the formula $C^2 = 3RT/m$ yields for $T = 400°$ K and $m = 40$ a mean square-root molecular velocity $C = 500$ m/sec, whereas the velocity of escape from the Moon is 2,380 m/sec. Even for $m = 10$, $C = 1,000$ m/sec only.*

Jeans[10] gives the following table of mean square-root molecular velocities:

Gas	at −100° C	O° C	300° C
H_2	1.47×10^5	1.84×10^5	2.66×10^5 cm/sec
He	1.04×10^5	1.31×10^5	1.90×10^5
H_2O	4.9×10^4	6.1×10^4	8.8×10^4
N_2	3.9×10^4	4.9×10^4	7.1×10^4
O_2	3.7×10^4	4.6×10^4	6.7×10^4
A	3.3×10^4	4.1×10^4	5.9×10^4
CO_2	3.1×10^4	3.9×10^4	5.7×10^4

300° C = 573° K, and it will be seen that at this temperature C exceeds the velocity of escape only in the case of hydrogen. The velocities of atomic hydrogen and oxygen will be higher by the factor of $\sqrt{2} \sim 1.41$, which will make the latter 9.45×10^4 cm/sec or 945 m/sec. The mean square-root velocity of free electrons will be $\sqrt{1840} \sim 43$ greater than that of atomic hydrogen, or $2.66 \times 1.41 \times 43 \times 10^5 \sim 1.613 \times 10^7$ cm/sec, which is nearly 7 times the escape velocity. They will be lost to space in a matter of hours.

The condition laid down by Elsmore[2] that 'the mean vertical component of velocity of the ionized gases is equal to the velocity of escape . . .' or 'appreciably greater than the escape velocity' (p. 1042) could apply only to an atmosphere of atomic hydrogen. It is beyond question that the Moon cannot permanently retain such an atmosphere. In other words, the argument amounts to assuming from the start that the Moon has no atmosphere and then showing that this non-existent atmosphere is very thin!

*Dr. Elsmore has gallantly acknowledged this in a private communication of 19.6.1969.

The case of argon of $m = 40$ is of some interest, as it may be (p. 218) the main constituent of the lunar atmosphere. I have, therefore, calculated the Jeans lifetime of an argon atmosphere from formula (4):

$$t_1 = \frac{C^3}{2g^2 a} \exp\left(\frac{3ga}{C^2}\right),$$

taking $C = 5 \times 10^4$ cm/sec (for $T = 400°$ K), $g = 160$ cm/sec^2 and $a = 1.75 \times 10^8$ cm. This yields t $= 1.730 \times 10^8$ years. It will be recalled, however, that t_1 is not the time in which the ground density of the atmosphere drops to 0, but only to $\frac{1}{e}$ of its initial value. $e^{4.5} \sim 100$, so that $4\frac{1}{2}$ such cycles will be needed to bring the atmospheric density down to 0.01 of the original figure, and this amounts to 778.5 million years. On the other hand for $m = 10$ the corresponding period is barely $25\frac{1}{2}$ days.

All doubts about the validity of Jeans's formula apart, this result is important for understanding the problem of lunar rivers. A predominantly argon atmosphere with a cold trap for water vapour and an ozonosphere evolved from the photo-dissociation of the latter (p. 211) could shield against loss to space even a comparatively modest quantity of surface water for some hundreds of millions of years.

Chapter 16

A perfectly black body reflects none of the incident radiation and re-emits all of the energy it has absorbed. This paragon can be fairly matched by the behaviour of a small hole in a hollow sphere whose inner surface has been lined with soot. The radiation issuing from such a hole is a *blackbody radiation*. Its energy E observes the Stefan-Boltzmann Law, whereby

$$E = \sigma T^4 , \qquad (9)$$

where σ is the Stefan-Boltzmann Constant equal to $(5.6692 \pm 0.0007) \times 10^{-5}$ ergs/cm^2/sec/deg., and T is the absolute temperature.

$$(9) \qquad\qquad T = \sqrt[4]{\frac{E}{\sigma}}. \qquad (10)$$

The behaviour of most substances departs considerably from this theoretical model, especially if strong colour is present. In most cases, however, a good approximation to reality can be obtained by recourse to a further fiction in the form of the so-called *perfectly grey body*; such a body observes equations (9) and (10) in respect of the absorbed fraction of the incident radiation, defined by the albedo. If the energy of the total incident radiation E is taken as unity the albedo A defines that fraction of it which is reflected; the absorbed fraction will, therefore, be $E (1 - A)$, and this expression should be substituted for E in (9) and (10) in the case of a grey body. The same applies to the further laws derived from (9).

Planck's Law[1] determines the intensity or brightness b_ν of the blackbody radiation of the frequency ν, which is the reciprocal of the wavelength λ multiplied into the velocity of light c, for a given absolute temperature T:

$$b_\nu = \frac{2h\nu^3 c^{-2}}{e^{h\nu/kT}-1}, \tag{11}$$

where e is the base of natural logarithms, h is Planck's Constant $= (6.6252 \pm 0.0002) \times 10^{-27}$ erg. sec., k Boltzmann's Constant $= (1.38046 \pm 0.00006) \times 10^{-16}$ erg. deg.$^{-1}$.

For a perfectly grey body the temperature T will be obtained from (10) after the correction for the albedo.

The wavelength of maximum intensity λ_{max} of the blackbody radiation at the absolute temperature T is given by Wien's Law[1]:

$$T \lambda_{max} = 0.28978 \text{ cm deg.} \tag{12}$$

In the case of visual or infrared radiation the situation is straight-forward enough, although it could be complicated by luminescence; but radio waves, including microwaves, may originate wholly or partly in the movement of electrical charges unrelated to temperature. In the case of the lunar radio emission it is assumed that any contribution from such sources is unimportant and the radiation is substantially thermal. In general, however, the temperature obtained from radio measurements on the assumption that it is thermal in origin and the emitter behaves as a perfectly black body is only a *brightness temperature*, which is a theoretical concept and need not bear any close relation to the actual temperature of the radiator. If, though, the intensity distribution over the spectrum of different frequencies complies with Planck's and Wien's Laws it may be inferred with reasonable assurance that the radiation is thermal.[14]

P. 233.

The thermal behaviour of the Moon during lunar eclipses is interpreted in terms of the product $k\rho c$, where k is the thermal conductivity, ρ the density and c the specific heat of the surface material.

From Pettit's[12] eclipse data Wesselink[1] obtained $(kc)^{\frac{1}{2}} = 0.0011$ c.g.s. units, which with $c = 0.2$ cal/g/deg., typical of common rocks, yields an extremely low thermal conductivity $k = 6 \times 10^{-6}$ cal/cm/sec/deg.

If now ρ is put at 2.0 g/cm³ we get

$$(k\rho c)^{-\frac{1}{2}} \sim 640. \tag{13}$$

The value of (13) is 20 for granite and basalt and 100 for pumice, but it becomes 1000 for fine powders under vacuum on the basis of the experiments carried out under very low pressures by the Polish physicist Smoluchowski in 1910. Hence Wesselink's conclusion that the Moon is covered with fine dust *in vacuo*.

As stated, the conductivity of particulate materials, such as dust, will depend on the pressure at contact between the grains, which is a function of gravity and will be *ceteris paribus* 6 times less on the Moon[6]. When an adjustment is made the conductivity of fine dust *in vacuo* will give too high a value for expression (13). On the other hand, this critically depends on ρ, which has been put at 2 and to the square-root of which (13) is inversely proportional. Thus if ρ is reduced to 0.2, as would be the case of a putative lunar reticulite, the product kc can be increased by $\sqrt{10} \sim 3.16$ and correspondingly higher k and/or c will satisfy Wesselink's condition.

Variability of k and c with temperature provides a further complication, not to mention the possible intervention of sorption (p. 241). Jaeger and Harper have found (1950)[9] that both Pettit's and Piddington and Minnett's data can be satisfied by the condition $(k\rho c)^{-\frac{1}{2}} = 1030$ c.g.s. units, but Muncey thought this need be no higher than 200 or 300 at 300° K,[3] which could be easily met by reticulite lavas.

There is an air of unreality about the whole discussion, but one thing is clear, the situation demands the surface formations being composed either of 'underdense' or finely divided incoherent material, or some combination of both. The *Surveyor* experiments have shown the surface material, though finely divided, is coherent; in fact, it has a kind of fairy-castle structure, though harder rocks are present. Now if such material had a bulk density as high as 1.5 g/cm³ or more it is very difficult to see how it could satisfy the thermal requirements.

REFERENCES

1 ALLEN, C. W. (1963). *Astrophysical Quantities*. 2nd Edn. Athlone Press, London.

2 ELSMORE, B. (1957). 'Radio Observations of the Lunar Atmosphere', *Philosophical Magazine*, **2**, p. 1,040 ff.

3 FIELDER, GILBERT (1961). *Structure of the Lunar Surface*. Pergamon Press, Oxford.

4 FIELDER, GILBERT (1965). *Lunar Geology*. Lutterworth Press, London.

5 FIRSOFF, V. A. (1959). *Science,* **130**, p. 1,337 f.

6 FIRSOFF, V. A. (1959). *Strange World of the Moon*. Hutchinson, London, and 1960, Basic Books, New York.

7   FIRSOFF, V. A. (1960). *Science*, **131** (3414), p. 1,669 f.

8   HOLMES, ARTHUR (1965). *Principles of Physical Geology*. Revised
    Edn. Nelson, Edinburgh and London.

9   JAEGER, J. C. (1959). 'Sub-surface temperatures on the Moon',
    Nature, **183**, p. 1,316 f.

10 JEANS, J. H. (1925). *The Dynamic Theory of Gases*. Cambridge
    University Press.

11 O'KEEFE, J. A., LOWMAN, P. D., JR., and CAMERON, W. S.
    (1967). 'Lunar Ring Dikes from Lunar Orbiter I', *Science*, **155**
    (3758), p. 77 ff.

12 PETTIT, EDISON (1940). Radiation Measurements on the Eclipsed
    Moon', *Astrophysical Journal*, **91**. p, 408 ff.

13 PIDDINGTON, J. C., and MINNETT, H. C. (1949). 'Microwave Radi-
    ation from the Moon', *Australian Journal of Scientific Research*,
    Series A, **2** (1).

14 PIDDINGTON, J. H. (1961). *Radio Astronomy*. Hutchinson, London.

15 SPITZER, LYMAN, JR. (1952) in KUIPER, G. P. (Ed.), *The Atmo-
    spheres of the Earth and Planets*, 2nd Edn. University of Chicago
    Press.

16 VAN DE HULST, H. C. (1954) in KUIPER, G. P. (Ed.), *The Sun*,
    p. 306 ff. University of Chicago Press.

17 WESSELINK, A. J. (1948). Heat Conductivity and Nature of the
    Lunar Surface Material', *Bulletin of the Astronomical Institutes of
    the Netherlands*, **10** (390), p. 35 ff.

Postscriptum

By Patrick Moore, O.B.E., F.R.A.S.

A postscript to a book written by another author usually indicates a considerable degree of eminence. This does not apply in the present case, since Firsoff's opinions carry much more weight than mine. However, since I have been so deeply involved in lunar controversy, it may be appropriate for me to say some things that he might be reluctant to say for himself.

On 21 July 1969, men landed on the Moon. First Neil Armstrong, then Edwin Aldrin stepped out on to the barren plain of the Sea of Tranquillity – and the dream of space-travel, which had been in the minds of pioneers for so many centuries, was a dream no longer. Various cherished theories were disproved on the spot. The deep-dust theory, which had nothing to recommend it from an observational point of view but which had exercised a strange grip in the minds of many, was finally killed. So was the idea that walking on the Moon would be impossible. With regard to the nature of the surface, more caution was necessary; and at the time when I write these words (1 August 1969) it still is. All that can really be said is that the results so far available are highly favourable to the theories put forward by Firsoff since the mid-1950s. Originally, his view was very much in the minority – and criticism was strong; as a personal recollection, I may add that I too came in for strong criticism when I published my first book and my first papers, around 1949, maintaining that vulcanism and not meteoritic impact had played the major rôle in the moulding of the Moon's surface.

To recapitulate briefly: Apollo 11, carrying three astronauts, was launched on 16 July 1969. It was put into a circum-lunar orbit in what has now come to be regarded as the conventional manner, and on the 21st, Armstrong and Aldrin, in their lunar module "Eagle", made a successful descent on to the surface, landing not far from the crater Möltke in the Mare Tranquillitatis. They spent less than three hours outside the module, but during this time they set up various experiments and also collected samples of lunar material. Of the experiments, the most interesting was in connection with a seismometer, which was scheduled to send back reports of any ground tremors which might be interpreted as moonquakes. Before long, a tremor was recorded which seems certainly to have been caused by movements in the Moon itself (and not by distortions in the abandoned lower stage of the module, which confused the record for some hours after Armstrong and Aldrin had blasted back into orbit to rendezvous with their colleague, Collins, who had remained circling the Moon at a height of about 70 miles in the command module of Apollo). The consensus of opinion at present is that the Moon retains considerable subcrustal heat, that it is layered, and that the crust has a thickness of about 20 kilometres. Moreover, a transient phenomenon in the crater Aristarchus, seen by the astronauts while they were going round the Moon on 19 July, was also observed by several ground-based observers of L.I.O.N. (Lunar International Observers' Network), including T. Moseley at Armagh Observatory. It now seems overwhelmingly probable that the Moon is not so inert as many people used to think.

Let me stress again that the question of the origin of the Moon's surface features has not been finally solved. Moreover, there are many modifications of many theories; even those who are firm in their beliefs that vulcanism has played the main part, differ in detail. But it certainly looks as though Firsoff has been right, for at least most of the time, when many other people were wrong; and all we can do now is to await results of the rock analysis at present being carried out. This will be followed by further landings, and we may be modestly confident that by the end of, say, 1970 we shall really have reliable information about the Moon world and its past history.

As an aside, it may be of historical interest to note that the present book, *The Old Moon and the New*, must be the last in which the text was actually written before the epic landing of Armstrong and Aldrin. It is also interesting to find that there are remarkably few revisions to be made as a result of the flight of Apollo 11. Of course, it may well be that revisions will be needed within the next few years; but this lies in the future. Finally, it is quite on the cards that some of the younger readers of this book will actually go to the Moon. If so, then they will at least have a good idea of what to expect upon what Firsoff has so aptly termed this "strange world".

Selsey,   PATRICK MOORE
1 *August* 1969

Index